T0258069

Encyclopedia of Hydrodynamics: Advanced Concepts

Volume IV

Encyclopedia of Hydrodynamics: Advanced Concepts
Volume IV

Edited by **Fay McGuire**

New York

Published by NY Research Press,
23 West, 55th Street, Suite 816,
New York, NY 10019, USA
www.nyresearchpress.com

Encyclopedia of Hydrodynamics: Advanced Concepts
Volume IV
Edited by Fay McGuire

International Standard Book Number: 978-1-63238-136-1 (Hardback)

This book contains information obtained from authentic and highly regarded sources. Copyright for all individual chapters remain with the respective authors as indicated. A wide variety of references are listed. Permission and sources are indicated; for detailed attributions, please refer to the permissions page. Reasonable efforts have been made to publish reliable data and information, but the authors, editors and publisher cannot assume any responsibility for the validity of all materials or the consequences of their use.

The publisher's policy is to use permanent paper from mills that operate a sustainable forestry policy. Furthermore, the publisher ensures that the text paper and cover boards used have met acceptable environmental accreditation standards.

Trademark Notice: Registered trademark of products or corporate names are used only for explanation and identification without intent to infringe.

Printed in the United States of America.

Contents

reface

is often said that books are a boon to mankind. They document every progress and pass the knowledge from one generation to the other. They play a crucial role in our lives. us I was both excited and nervous while editing this book. I was pleased by the thought being able to make a mark but I was also nervous to do it right because the future of dents depends upon it. Hence, I took a few months to research further into the discipline, vise my knowledge and also explore some more aspects. Post this process, I begun with e editing of this book.

is book examines novel viewpoints about procedures and tools used in Hydrodynamics. e phenomena associated with the flow of fluids are usually complex, and tough to quantify. ovel approaches - considering points of view still not investigated - may present useful vices in the study of hydrodynamics and the associated transport phenomenon. The ecifications of the flows and the characteristics of the fluids must be studied on a small ale. Subsequently, novel concepts and devices are devised to better explain the fluids and eir characteristics. This book provides conclusions about advanced issues of calculated and served flows. Major topics in this book are radiation, electro-magneto-hydrodynamics and agneto-rheology; special points on simulations and experimental inputs are also discussed.

thank my publisher with all my heart for considering me worthy of this unparalleled portunity and for showing unwavering faith in my skills. I would also like to thank e editorial team who worked closely with me at every step and contributed immensely wards the successful completion of this book. Last but not the least, I wish to thank my iends and colleagues for their support.

Editor

Part 1

Radiation-, Electro-, Magnetohydrodynamics and Magnetorheology

An IMEX Method for the Euler Equations That Posses Strong Non-Linear Heat Conduction and Stiff Source Terms (Radiation Hydrodynamics)

Samet Y. Kadioglu[1] and Dana A. Knoll[2]

[1]*Idaho National Laboratory, Fuels Modeling and Simulation Department, Idaho Falls*
[2]*Los Alamos National Laboratory, Theoretical Division, Los Alamos*
USA

1. Introduction

Here, we present a truly second order time accurate self-consistent IMEX (IMplicit/EXplicit) method for solving the Euler equations that posses strong nonlinear heat conduction and very stiff source terms (Radiation hydrodynamics). This study essentially summarizes our previous and current research related to this subject (Kadioglu & Knoll, 2010; 2011; Kadioglu, Knoll & Lowrie, 2010; Kadioglu, Knoll, Lowrie & Rauenzahn, 2010; Kadioglu et al., 2009; Kadioglu, Knoll, Sussman & Martineau, 2010). Implicit/Explicit (IMEX) time integration techniques are commonly used in science and engineering applications (Ascher et al., 1997; 1995; Bates et al., 2001; Kadioglu & Knoll, 2010; 2011; Kadioglu, Knoll, Lowrie & Rauenzahn, 2010; Kadioglu et al., 2009; Khan & Liu, 1994; Kim & Moin, 1985; Lowrie et al., 1999; Ruuth, 1995). These methods are particularly attractive when dealing with physical systems that consist of multiple physics (multi-physics problems such as coupling of neutron dynamics to thermal-hydrolic or to thermal-mechanics in reactors) or fluid dynamics problems that exhibit multiple time scales such as advection-diffusion, reaction-diffusion, or advection-diffusion-reaction problems. In general, governing equations for these kinds of systems consist of stiff and non-stiff terms. This poses numerical challenges in regards to time integrations, since most of the temporal numerical methods are designed specific for either stiff or non-stiff problems. Numerical methods that can handle both physical behaviors are often referred to as IMEX methods. A typical IMEX method isolates the stiff and non-stiff parts of the governing system and employs an explicit discretization strategy that solves the non-stiff part and an implicit technique that solves the stiff part of the problem. This standard IMEX approach can be summarized by considering a simple prototype model. Let us consider the following scalar model

$$u_t = f(u) + g(u), \tag{1}$$

where $f(u)$ and $g(u)$ represent non-stiff and stiff terms respectively. Then the IMEX strategy consists of the following algorithm blocks:
Explicit block solves:

$$\frac{u^* - u^n}{\Delta t} = f(u^n), \tag{2}$$

Implicit block solves:

$$\frac{u^{n+1} - u^*}{\Delta t} = g(u^{n+1}). \tag{3}$$

Here, for illustrative purposes we used only first order time differencing. In literature although the both algorithm blocks are formally written as second order time discretizations, the classic IMEX methods (Ascher et al., 1997; 1995; Bates et al., 2001; Kim & Moin, 1985; Lowrie et al., 1999; Ruuth, 1995) split the operators in such a way that the implicit and explicit blocks are executed independent of each other resulting in non-converged non-linearities therefore time inaccuracies (order reduction to first order is often reported for certain applications). Below, we illustrate the interaction of an explicit and an implicit algorithm block based on second order time discretizations of Equation(1) in classical sense,
Explicit block:

$$u^1 = u^n + \Delta t f(u^n)$$
$$u^* = (u^1 + u^n)/2 + \Delta t/2 f(u^1) \tag{4}$$

Implicit block:

$$u^{n+1} = u^* + \Delta t/2[g(u^n) + g(u^{n+1})]. \tag{5}$$

Notice that the explicit block is based on a second order TVD Runge-Kutta method and the implicit block uses the Crank-Nicolson method (Gottlieb & Shu, 1998; LeVeque, 1998; Thomas, 1999). The major drawback of this strategy as mentioned above is that it does not preserve the formal second order time accuracy of the whole algorithm due to the absence of sufficient interactions between the two algorithm blocks (refer to highlighted terms in Equation (4)) (Bates et al., 2001; Kadioglu, Knoll & Lowrie, 2010).
In an alternative IMEX approach that we have studied extensively in (Kadioglu & Knoll, 2010; 2011; Kadioglu, Knoll & Lowrie, 2010; Kadioglu, Knoll, Lowrie & Rauenzahn, 2010; Kadioglu et al., 2009), the explicit block is always solved inside the implicit block as part of the nonlinear function evaluation making use of the well-known Jacobian-Free Newton Krylov (JFNK) method (Brown & Saad, 1990; Knoll & Keyes, 2004). We refer this IMEX approach as *a self-consistent IMEX method*. In this strategy, there is a continuous interaction between the implicit and explicit blocks meaning that the improved solutions (in terms of time accuracy) at each nonlinear iteration are immediately felt by the explicit block and the improved explicit solutions are readily available to form the next set of nonlinear residuals. This continuous interaction between the two algorithm blocks results in an implicitly balanced algorithm in that all nonlinearities due to coupling of different time terms are consistently converged. In other words, we obtain an IMEX method that eliminates potential order reductions in time accuracy (the formal second order time accuracy of the whole algorithm is preserved). Below, we illustrate the interaction of the explicit and implicit blocks of the self-consistent IMEX method for the scalar model in Equation (1). The interaction occurs through the highlighted terms in Equation (6).
Explicit block:

$$u^1 = u^n + \Delta t f(u^n)$$
$$u^* = (u^1 + u^n)/2 + \Delta t/2 f(u^{n+1}) \tag{6}$$

Implicit block:

$$u^{n+1} = u^* + \Delta t/2[g(u^n) + g(u^{n+1})]. \tag{7}$$

Remark: We remark that another way of achieving a self-consistent IMEX integration that preserves the formal numerical accuracy of the whole system is to improve the lack of influence of the explicit and implicit blocks on one another by introducing an external iteration procedure wrapped around the both blocks. More details regarding this methodology can be found in (Kadioglu et al., 2005).

. Applications

We have applied the above described self-consistent IMEX method to both multi-physics and multiple time scale fluid dynamics problems (Kadioglu & Knoll, 2010; 2011; Kadioglu, Knoll, Lowrie & Rauenzahn, 2010; Kadioglu et al., 2009; Kadioglu, Knoll, Sussman & Martineau, 2010). The multi-physics application comes from a multi-physics analysis of fast burst reactor study (Kadioglu et al., 2009). The model couples a neutron dynamics that simulates the transient behavior of neutron populations to a mechanics model that predicts material expansions and contractions. It is important to introduce a second order accurate numerical procedure for this kind of nonlinearly coupled system, because the criticality and safety study can depend on how well we predict the feedback between the neutronics and the mechanics of the fuel assembly inside the reactor. In our second order self-consistent IMEX framework, the mechanics part is solved explicitly inside the implicit neutron diffusion block as part of the nonlinear function evaluation. We have reported fully second order time convergent calculations for this model (Kadioglu et al., 2009).

As part of the multi-scale fluid dynamics application, we have solved multi-phase flow problems which are modeled by incompressible two-phase Navier-Stokes equations that govern the flow dynamics plus a level set equation that solves the inter-facial dynamics between the fluids (Kadioglu, Knoll, Sussman & Martineau, 2010). In these kinds of models, there is a strong non-linear coupling between the interface and fluid dynamics, e.g, the viscosity coefficient and surface tension forces are highly non-linear functions of interface variables, on the other hand, the fluid interfaces are advected by the flow velocity. Therefore, it is important to introduce an accurate integration technique that converges all non-linearities due to the strong coupling. Our self-consistent IMEX method operates on this model as follows; the interface equation together with the hyperbolic parts of the fluid equations are treated explicitly and solved inside an implicit loop that solves the viscous plus stiff surface tension forces. More details about the splitting of the operators of the Navier-Stokes equations in a self-consistent IMEX manner can be found in (Kadioglu & Knoll, 2011).

Another multi-scale fluid dynamics application comes from radiation hydrodynamics that we will be focusing on in the remainder of this chapter. Radiation hydrodynamics models are commonly used in astrophysics, inertial confinement fusion, and other high-temperature flow systems (Bates et al., 2001; Castor, 2006; Dai & Woodward, 1998; Drake, 2007; Ensman, 1994; Kadioglu & Knoll, 2010; Lowrie & Edwards, 2008; Lowrie & Rauenzahn, 2007; Mihalas & Mihalas, 1984; Pomraning, 1973). A commonly used model considers the compressible Euler equations that contains a non-linear heat conduction term in the energy part. This model is relatively simple and often referred to as a *Low Energy-Density Radiation Hydrodynamics (LERH)* in a diffusion approximation limit (Kadioglu & Knoll, 2010). A more complicated model is referred to as a *High Energy-Density Radiation Hydrodynamics (HERH)* in a diffusion approximation limit that considers a combination of a hydrodynamical model resembling the compressible Euler equations and a radiation energy model that contains a separate radiation energy equation with nonlinear diffusion plus coupling source terms to materials (Kadioglu, Knoll, Lowrie & Rauenzahn, 2010). Radiation Hydrodynamics problems are difficult to tackle numerically since they exhibit multiple time scales. For instance, radiation and hydrodynamics process can occur on time scales that can differ from each other by many orders of magnitudes. Hybrid methods (Implicit/Explicit (IMEX) methods) are highly desirable for these kinds of models, because if one uses all explicit discretizations then due to very stiff diffusion process the explicit time steps become often impractically small to satisfy stability conditions (LeVeque, 1998; Thomas, 1999). Previous IMEX attempts to solve these problems were not quite successful, since they often reported order reductions in time accuracy (Bates et al., 2001; Lowrie et al., 1999). The main reason for time inaccuracies was how the explicit and implicit operators were split in which explicit solutions were lagging behind the implicit ones. In our self-consistent IMEX method, the hydrodynamics part is solved explicitly making use of the well-understood explicit schemes within an implicit diffusion block that corresponds to radiation transport. Explicit solutions are obtained as part of the non-linear functions evaluations withing the JFNK framework. This strategy has enabled us to produce fully second order time accurate results for both LERH and more complicated HERH models (Kadioglu & Knoll, 2010; Kadioglu, Knoll, Lowrie & Rauenzahn, 2010).

In the following sections, we will go over more details about the LERH and HERH models and the implementation/implications of the self-consistent IMEX technology when it is applied to these models. We will also present a mathematical analysis that reveals the analytical convergence behavior of our method and compares it to a classic IMEX approach.

2.1 A Low Energy Density Radiation Hydrodynamics Model (LERH)

This model uses the following system of partial differential equations formulated in spherically symmetric coordinates.

$$\frac{\partial \rho}{\partial t} + \frac{1}{r^2}\frac{\partial}{\partial r}(r^2 \rho u) = 0, \tag{8}$$

$$\frac{\partial}{\partial t}(\rho u) + \frac{1}{r^2}\frac{\partial}{\partial r}(r^2 \rho u^2) + \frac{\partial p}{\partial r} = 0, \tag{9}$$

$$\frac{\partial E}{\partial t} + \frac{1}{r^2}\frac{\partial}{\partial r}[r^2 u(E + p)] = \frac{1}{r^2}\frac{\partial}{\partial r}(r^2 \kappa \frac{\partial T}{\partial r}), \tag{10}$$

where ρ, u, p, E, and T are the mass density, flow velocity, fluid pressure, total energy density of the fluid, and the fluid temperature respectively. κ is the coefficient of thermal conduction (or diffusion coefficient) and in general is a nonlinear function of ρ and T. In this study, we will use an ideal gas equation of state, i.e, $p = R\rho T = (\gamma - 1)\rho e$, where R is the specific gas constant per unit mass, γ is the ratio of specific heats, and e is the internal energy of the fluid per unit mass. The coefficient of thermal conduction will be assumed to be written as a power law in density and temperature, i.e, $\kappa = \kappa_0 \rho^a T^b$, where κ_0, a and b are constants (Marshak, 1958). This simplified radiation hydrodynamics model allows one to study the dynamics of nonlinearly coupled two distinct physics; compressible fluid flow and nonlinear diffusion.

2.2 A High Energy Density Radiation Hydrodynamics Model (HERH)

In general, the radiation hydrodynamics concerns the propagation of thermal radiation through a fluid and the effect of this radiation on the hydrodynamics describing the fluid motion. The role of the thermal radiation increases as the temperature is raised. At low emperatures the radiation effects are negligible, therefore, a low energy density model LERH) that limits the radiation effects to a non-linear heat conduction is sufficient. However, at high temperatures, a more complicated high energy density radiation hydrodynamics HERH) model that accounts for more significant radiation effects has to be considered. Accordingly, the governing equations of the HERH model consist of the following system

$$\frac{\partial \rho}{\partial t} + \frac{1}{r^2}\frac{\partial}{\partial r}(r^2\rho u) = 0, \tag{11}$$

$$\frac{\partial}{\partial t}(\rho u) + \frac{1}{r^2}\frac{\partial}{\partial r}(r^2\rho u^2) + \frac{\partial}{\partial r}(p + p_\nu) = 0, \tag{12}$$

$$\frac{\partial E}{\partial t} + \frac{1}{r^2}\frac{\partial}{\partial r}[r^2 u(E+p)] = -c\sigma_a(aT^4 - E_\nu) - \frac{1}{3}u\frac{\partial E_\nu}{\partial r}, \tag{13}$$

$$\frac{\partial E_\nu}{\partial t} + \frac{1}{r^2}\frac{\partial}{\partial r}[r^2 u(E_\nu + p_\nu)] = \frac{1}{r^2}\frac{\partial}{\partial r}(r^2 cD_r\frac{\partial E_\nu}{\partial r}) + c\sigma_a(aT^4 - E_\nu) + \frac{1}{3}u\frac{\partial E_\nu}{\partial r}, \tag{14}$$

where the flow variables and parameters that also occur in the LERH model are described above. Here, more variable definitions come from the radiation physics, i.e, E_ν is the radiation energy density, $p_\nu = \frac{E_\nu}{3}$ is the radiation pressure, c is the speed of light, a is the Stephan-Boltzmann constant, σ_a is the macroscopic absorption cross-section, and D_r is the radiation diffusion coefficient. From the simple diffusion theory, D_r can be written as

$$D_r(T) = \frac{1}{3\sigma_a}. \tag{15}$$

We note that we solve a non-dimensional version of Equations (11)-(14) in order to normalize large digit numbers (c, σ_a, a etc.) and therefore improve the performance of the non-linear solver. The details of the non-dimensionalization procedure are given in (Kadioglu, Knoll, Lowrie & Rauenzahn, 2010). The non-dimensional system is the following,

$$\frac{\partial \rho}{\partial t} + \frac{1}{r^2}\frac{\partial}{\partial r}(r^2\rho u) = 0, \tag{16}$$

$$\frac{\partial}{\partial t}(\rho u) + \frac{1}{r^2}\frac{\partial}{\partial r}(r^2\rho u^2) + \frac{\partial}{\partial r}(p + \mathcal{P}p_\nu) = 0, \tag{17}$$

$$\frac{\partial E}{\partial t} + \frac{1}{r^2}\frac{\partial}{\partial r}[r^2 u(E+p)] = -\mathcal{P}\sigma_a(T^4 - E_\nu) - \frac{1}{3}\mathcal{P}u\frac{\partial E_\nu}{\partial r}, \tag{18}$$

$$\frac{\partial E_\nu}{\partial t} + \frac{1}{r^2}\frac{\partial}{\partial r}[r^2 u(E_\nu + p_\nu)] = \frac{1}{r^2}\frac{\partial}{\partial r}(r^2\kappa\frac{\partial E_\nu}{\partial r}) + \sigma_a(T^4 - E_\nu) + \frac{1}{3}u\frac{\partial E_\nu}{\partial r}, \tag{19}$$

where $\mathcal{P} = \frac{aT_0^4}{\rho_0 c_{s,0}^2}$ is a non-dimensional parameter that measures the radiation effects on the flow and is roughly proportional to the ratio of the radiation and fluid pressures.

3. Numerical procedure

Here, we present the numerical procedure for the LERH model. The extension to the HERH model is straight forward. First, we split the operators of Equations (8)-(10) into two pieces one being the pure hydrodynamics part (hyperbolic conservation laws) and the other

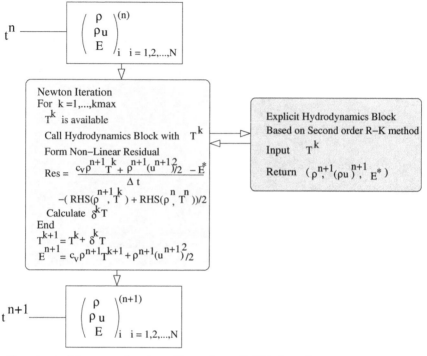

Fig. 1. Flowchart of the second order self-consistent IMEX algorithm

accounting for the effects of radiation transport (diffusion equation). For instance, the pure hydrodynamics equations can be written as

$$\frac{\partial \mathbf{U}}{\partial t} + \frac{\partial (A\mathbf{F})}{\partial V} + \frac{\partial \mathbf{G}}{\partial r} = 0, \tag{20}$$

where $\mathbf{U} = (\rho, \rho u, E)^T$, $\mathbf{F}(\mathbf{U}) = (\rho u, \rho u^2, u(E + p))^T$, and $\mathbf{G}(\mathbf{U}) = (0, p, 0)^T$. Then the diffusion equation becomes

$$\frac{\partial E}{\partial t} = \frac{\partial}{\partial V}(A\kappa \frac{\partial T}{\partial r}), \tag{21}$$

where $V = \frac{4}{3}\pi r^3$ is the generalized volume coordinate in one-dimensional spherical geometry, and $A = 4\pi r^2$ is the associated cross-sectional area. Notice that the total energy density, E, obtained by Equation (20) just represents the hydrodynamics component and it must be augmented by Equation (21).

Our algorithm consists of an explicit and an implicit block. The explicit block solves Equation (20) and the implicit block solves Equation (21). We will briefly describe these algorithm blocks in the following subsections. However, we note again that the explicit block is embedded within the implicit block as part of a nonlinear function evaluation as it is depicted in Fig. 1. This is done to obtain a nonlinearly converged algorithm that leads to second order calculations. We also note that similar discretizations, but without converging nonlinearities, can lead to order reduction in time convergence (Bates et al., 2001). Before we go into details of the individual algorithm blocks, we would like to present a flow diagram that illustrates the

xecution of the whole algorithm in the self-consistent IMEX sense (refer to Fig. 1). According
o this diagram, at beginning of each Newton iteration, we have the temperature values based
n the current Newton iterate. This temperature is passed to the explicit block that returns the
ıpdated density, momentum, and a prediction to total energy. Then we form the non-linear
esiduals (e.g, forming the IMEX function in Section 3.3) for the diffusion equation out of
he updated and predicted values. With the IMEX function in hand, we can execute the JFNK
method. After the Newton method convergences, we get second order converged temperature
ınd total energy density field.

ı.1 Explicit block

Jur explicit time discretization is based on a second order TVD Runge-Kutta method
Gottlieb & Shu, 1998; Gottlieb et al., 2001; Shu & Osher, 1988; 1989). The main reason why we
choose this methodology is that it preserves the strong stability properties of the explicit Euler
method. This is important because it is well known that solutions to the conservation laws
ısually involve discontinuities (e.g, shock or contact discontinuities) and (Gottlieb & Shu,
998; Gottlieb et al., 2001) suggest that a time integration method which has the strong
tability preserving property leads to non-oscillatory calculations (especially at shock or
contact discontinuities).
A second order two-step TVD Runge-Kutta method for (20) can be cast as
Step-1 :

$$\rho^1 = \rho^n - \Delta t \frac{1}{r^2} \frac{\partial}{\partial r} (r^2 \rho u)^n,$$

$$(\rho u)^1 = (\rho u)^n - \Delta t [\frac{1}{r^2} \frac{\partial}{\partial r} (r^2 \rho u^2) + \frac{\partial p}{\partial r}]^n,$$

$$E^1 = E^n - \Delta t \{\frac{1}{r^2} \frac{\partial}{\partial r} [r^2 u (E + p)]\}^n,$$

$$(22)$$

Step-2 :

$$\rho^{n+1} = \frac{\rho^n + \rho^1}{2} - \frac{\Delta t}{2} \frac{1}{r^2} \frac{\partial}{\partial r} (r^2 \rho u)^1,$$

$$(\rho u)^{n+1} = \frac{(\rho u)^n + (\rho u)^1}{2} - \frac{\Delta t}{2} \{\frac{1}{r^2} \frac{\partial}{\partial r} (r^2 \rho u^2)^1 + \frac{\partial}{\partial r} (\rho^1 R T^{n+1})\},$$

$$E^* = \frac{E^n + E^1}{2} - \frac{\Delta t}{2} \{\frac{1}{r^2} \frac{\partial}{\partial r} [r^2 u^1 (c_v \rho^1 T^{n+1} + \frac{1}{2} \rho^1 (u^1)^2 + \rho^1 R T^{n+1})]\}.$$

$$(23)$$

Ne used the following equation of state relations in (22)- (23);

$$p = \rho R T E = c_v \rho T + \frac{1}{2} \rho u^2,$$

$$(24)$$

where $c_v = \frac{R}{\gamma - 1}$ is the fluid specific heat with R being the universal gas constant. This
explicit algorithm block interacts with the implicit block through the highlighted T^{n+1} terms
n Equation (23). We can observe that the implicit equation (21) is practically solved for T
ɔy using the energy relation. Therefore, the explicit block is continuously impacted by the

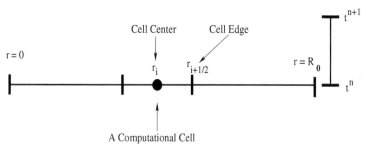

u_i^n : Represents a Cell Centered Quantity at time level n

$u_{i+1/2}^n$: Represents a Cell Edge Quantity at time level n

Fig. 2. Computational Conventions.

implicit T^{n+1} solutions at each non-linear Newton iteration. This provides the tight nonlinear coupling between the two algorithm blocks. Notice that the kth nonlinear Newton iteration of the implicit block corresponds to $T^{n+1} \leftarrow T^k$ and $k \rightarrow (n+1)$ upon the convergence of the Newton method (refer to Fig. 1). Also, the ∗ values in Equation (23) are predicted intermediate values and later they are corrected by the implicit block which is given in the next subsection. One observation about this algorithm block is that some calculations are redundant related to Equation (22). In other words, Equation (22) can be computed only once at the beginning of each Newton iteration, because the non-linear iterations do not impact (22). This can lead overall less number of function evaluations.

Now we shall describe how we evaluate the numerical fluxes needed by Equations (22) and (23). For simplicity, we consider (20) to describe our fluxing procedure. Basically, it is based on the Local Lax Friedrichs (LLF) method (we refer to (LeVeque, 1998; Thomas, 1999) for the details of the LLF method and for more information in regards to the explicit discretizations of conservation laws). For instance, if we consider the following simple discretization for Equation (20),

$$\mathbf{U}_i^1 = \mathbf{U}_i^n - \frac{\Delta t}{\Delta V_i}(A_{i+1/2}F_{i+1/2}^n - A_{i-1/2}F_{i-1/2}^n) - \frac{\Delta t}{\Delta r}(G_{i+1/2}^n - G_{i+1/2}^n), \qquad (25)$$

where $\Delta V_i = V(r_{i+1/2}) - V(r_{i-1/2})$, $A_{i\pm1/2} = A(r_{i\pm1/2})$, and indices i and $i+1/2$ represent cell center and cell edge values respectively (refer to Fig. 2), then the *Local Lax Friedrichs method* defines $F_{i+1/2}$ and $G_{i+1/2}$ as

$$F_{i+1/2} = \frac{F(\mathbf{U}_{i+1/2}^R) + F(\mathbf{U}_{i+1/2}^L)}{2} - \alpha_{i+1/2}\frac{\mathbf{U}_{i+1/2}^R - \mathbf{U}_{i+1/2}^L}{2}, \qquad (26)$$

$$G_{i+1/2} = \frac{G(\mathbf{U}_{i+1/2}^R) + G(\mathbf{U}_{i+1/2}^L)}{2}, \qquad (27)$$

where $\alpha = max\{|\lambda_1^L|, |\lambda_1^R|, |\lambda_2^L|, |\lambda_2^R|, |\lambda_3^L|, |\lambda_3^R|\}$ in which $\lambda_1 = u - c$, $\lambda_2 = u$, $\lambda_3 = u + c$, and c is the sound speed. The sound speed is defined by

$$c = \sqrt{\frac{\partial p}{\partial \rho}}, \qquad (28)$$

n IMEX Method for the Euler Equations That Posses Strong Non-Linear Heat Conduction and Stiff Source Terms
Radiation Hydrodynamics)

11

where $\frac{\partial p}{\partial \rho} = RT$ in this study. $\mathbf{U}^R_{i+1/2}$ and $\mathbf{U}^L_{i+1/2}$ are the interpolated values at $(i+1/2)^{th}$ cell edge from the right and left side, i.e,

$$\mathbf{U}^R_{i+1/2} = \mathbf{U}_{i+1} - \frac{\Delta r}{2}\mathbf{U}_{r,i+1},$$

$$\mathbf{U}^L_{i+1/2} = \mathbf{U}_i + \frac{\Delta r}{2}\mathbf{U}_{r,i}, \tag{29}$$

where

$$\mathbf{U}_{r,i} = minmod(a,b) = \begin{cases} a \text{ if } |a| < |b| \text{ and } ab > 0, \\ b \text{ if } |b| < |a| \text{ and } ab > 0, \\ 0 \text{ if } ab \le 0, \end{cases} \tag{30}$$

where

$$a = \frac{\mathbf{U}_{i+1} - \mathbf{U}_i}{\Delta r}, \tag{31}$$

$$b = \frac{\mathbf{U}_i - \mathbf{U}_{i-1}}{\Delta r}. \tag{32}$$

3.2 Implicit block

The explicit block produces the following solution vector

$$\mathbf{U}^n \rightarrow \mathbf{U}^* = \begin{pmatrix} \rho^{n+1} \\ (\rho u)^{n+1} \\ E^* \end{pmatrix}.$$

This information is used to discretize Equation (21) as follows,

$$\frac{(c_v \rho^{n+1} T^{n+1} + \frac{1}{2}\rho^{n+1}(u^{n+1})^2 - E^*)_i}{\Delta t} = \frac{1}{2}\frac{\partial}{\partial V}(A\kappa^{n+1}\frac{\partial T^{n+1}}{\partial r})_i + \frac{1}{2}\frac{\partial}{\partial V}(A\kappa^n\frac{\partial T^n}{\partial r})_i, \tag{33}$$

where

$$\frac{\partial}{\partial V}(A\kappa\frac{\partial T}{\partial r})_i = \frac{A_{i+1/2}\kappa_{i+1/2}(T_{i+1} - T_i)/\Delta r}{\Delta V_i} - \frac{A_{i-1/2}\kappa_{i-1/2}(T_i - T_{i-1})/\Delta r}{\Delta V_i}. \tag{34}$$

Notice that this implicit discretization resembles to the *Crank-Nicolson method* (Strikwerda, 1989; Thomas, 1998). We solve Equation (33) iteratively for T^{n+1}. The nonlinear solver needed by Equation (33) is based on the Jacobian-Free Newton Krylov method which is described in the next subsection. When the Newton method converges all the nonlinearities in this discretization, we obtain the following fully updated solution vector,

$$\mathbf{U}^* \rightarrow \mathbf{U}^{n+1} = \begin{pmatrix} \rho^{n+1} \\ (\rho u)^{n+1} \\ E^{n+1} \end{pmatrix}.$$

3.3 The Jacobian-Free Newton Krylov method and forming the IMEX function

The Jacobian-Free Newton Krylov method (e.g, refer to (Brown & Saad, 1990; Kelley, 2003; Knoll & Keyes, 2004)) is a combination of the Newton method that solves a system of nonlinear equations and a Krylov subspace method that solves the Newton correction equations. With this method, Newton-like super-linear convergence is achieved in the nonlinear iterations, without the complexity of forming or storing the Jacobian matrix. The effects of the Jacobian matrix are probed only through approximate matrix-vector products required in the Krylov iterations. Below, we provide more details about this technique.

The Newton method solves $\mathbf{F}(T) = 0$ (e.g, assume Equation (33) is written in this form) iteratively over a sequence of linear system defined by

$$\mathbf{J}(T^k)\delta T^k = -\mathbf{F}(T^k),$$
$$T^{k+1} = T^k + \delta T^k, \qquad k = 0, 1, \cdots \tag{35}$$

where $\mathbf{J}(T^k) = \frac{\partial \mathbf{F}}{\partial T}$ is the Jacobian matrix and δT^k is the update vector. The Newton iteration is terminated based on a required drop in the norm of the nonlinear residual, i.e,

$$\|\mathbf{F}(T^k)\|_2 < tol_{res}\|\mathbf{F}(T^0)\|_2 \tag{36}$$

where tol_{res} is a given tolerance. The linear system, Newton correction equation (35), is solved by using the Arnoldi based Generalized Minimal RESidual method (GMRES)(Saad, 2003) which belongs to the general class of the Krylov subspace methods(Reid, 1971). We note that these subspace methods are particularly suitable choice when dealing with non-symmetric linear systems. In GMRES, an initial linear residual, $\mathbf{r}_0$, is defined for a given initial guess δT_0,

$$\mathbf{r}_0 = -\mathbf{F}(T) - \mathbf{J}\delta T_0. \tag{37}$$

Here we dropped the index k convention since the Krylov (GMRES) iteration is performed at a fixed k. Let j be the Krylov iteration index. The jth Krylov iteration minimizes $\|\mathbf{J}\delta T_j + \mathbf{F}(T)\|_2$ within a subspace of small dimension, relative to n (the number of unknowns), in a least-squares sense. δT_j is drawn from the subspace spanned by the Krylov vectors, $\{\mathbf{r}_0, \mathbf{J}\mathbf{r}_0, \mathbf{J}^2\mathbf{r}_0, \cdots, \mathbf{J}^{j-1}\mathbf{r}_0\}$, and can be written as

$$\delta T_j = \delta T_0 + \sum_{i=0}^{j-1} \beta_i (\mathbf{J})^i \mathbf{r}_0, \tag{38}$$

where the scalar β_i minimizes the residual. The Krylov iteration is terminated based on the following inexact Newton criteria (Dembo, 1982)

$$\|\mathbf{J}\delta T_j + \mathbf{F}(T)\|_2 < \gamma\|\mathbf{F}(T)\|_2, \tag{39}$$

where the parameter γ is set in terms of how tight the linear solver should converge at each Newton iteration (we typically use $\gamma = 10^{-3}$). One particularly attractive feature of this methodology is that it does not require forming the Jacobian matrix. Instead, only matrix-vector multiplications, $\mathbf{J}v$, are needed, where $v \in \{\mathbf{r}_0, \mathbf{J}\mathbf{r}_0, \mathbf{J}^2\mathbf{r}_0, \cdots\}$. This leads to the so-called *Jacobian-Free* implementations in which the action of the Jacobian matrix can be approximated by

$$\mathbf{J}v = \frac{\mathbf{F}(T + \epsilon v) - \mathbf{F}(T)}{\epsilon}, \tag{40}$$

An IMEX Method for the Euler Equations That Posses Strong Non-Linear Heat Conduction and Stiff Source Terms
Radiation Hydrodynamics)

13

where $\epsilon = \frac{1}{n\|v\|_2} \sum_{i=1}^{n} b|u_i| + b$, n is the dimension of the linear system and b is a constant whose magnitude is within a few orders of magnitude of the square root of machine roundoff (typically 10^{-6} for 64-bit double precision).

Here, we briefly describe how to form the IMEX function $F(T)$. We refer $F(T)$ as the IMEX function, since it uses both explicit (hydrodynamics) and implicit (diffusion) information. Notice that for a method that uses all implicit information, $F(T)$ would correspond to a regular nonlinear residual function. The following pseudo code describes how to form $F(T)$ (we also refer to Fig. 1).

Evaluating $F(T^k)$:

Given T^k where k represents the current Newton iteration.
Call Hydrodynamics block with (ρ^n, u^n, E^n, T^k) to compute ρ^{n+1}, u^{n+1}, E^*.
Form $F(T^k)$ based on the Crank-Nicolson method,
$$F(T^k) = \frac{[c_v \rho^{n+1} T^k + \frac{1}{2}\rho^{n+1}(u^{n+1})^2 - E^*]}{\Delta t} - \frac{1}{2}\frac{\partial}{\partial V}\left(A\kappa^k \frac{\partial T^k}{\partial r}\right) - \frac{1}{2}\frac{\partial}{\partial V}\left(A\kappa^n \frac{\partial T^n}{\partial r}\right).$$

It is important to note that we are not iterating between the implicit and explicit blocks. Instead we are executing the explicit block inside of a nonlinear function evaluation defined by $F(T^k)$. The unique properties of JFNK allow us to perform a Newton iteration on this IMEX function, and thus JFNK is a required component of this nonlinearly converged IMEX approach.

3.4 Time step control
In this section, we describe two procedures to determine the computational time steps that are used in our test calculations. The first one was originally proposed by (Rider & Knoll, 1999). The idea is to estimate the dominant wave propagation speed in the problem. In one dimension this involves calculating the ratio of temporal to spatial derivatives of the dependent variables. In principle, it is sufficient to consider the following hyperbolic equation rather than using the entire system of the governing equations

$$\frac{\partial E}{\partial t} + v_f \frac{\partial E}{\partial r} = 0, \tag{41}$$

where the unknown v_f represents the front velocity. This gives

$$v_f = -\frac{\partial E/\partial t}{\partial E/\partial r}. \tag{42}$$

As noted in Rider & Knoll (1999), to avoid problems from lack of smoothness the following numerical approximation is used to calculate v_f

$$v_f^n = \frac{\sum(|E_i^n - E_i^{n-1}|/\Delta t)}{\sum(|E_{i+1}^n - E_{i-1}^n|/2\Delta r)}. \tag{43}$$

Then the new time step is determined by the Courant-Friedrichs-Lewy (CFL) condition

$$\Delta t^{n+1} = C\frac{\|\Delta r\|}{v_f^n}, \tag{44}$$

where $\| \Delta r \|$ uses the L_1 norm as in Equation (43). We can further simplify Equation (44) by using Equation (43), i.e,

$$\Delta t^{n+1} = \frac{1}{2} \frac{\sum |E_{i+1}^n - E_{i-1}^n|}{\sum (|E_i^n - E_i^{n-1}|/\Delta t)}. \tag{45}$$

We remark that the time steps determined by this procedure is always compared with the pure hydrodynamics time steps and the most restrictive ones are selected. The hydrodynamics time steps are calculated by

$$\Delta t^{Hydro,n+1} = CFL \times \frac{\Delta r}{max_i |u + c|_i}, \tag{46}$$

where u is the fluid velocity and c is the sound speed (e.g, refer to Equation (28)). The coefficient CFL is set to 0.5. Alternative time step control criterion are used for radiation hydrodynamics problems (Bowers & Wilson, 1991). One commonly used approach is based on monitoring the maximum relative change in E. For instance,

$$\Delta t^{n+1} = \Delta t^n \sqrt{\frac{(\Delta E/E)^{n+1}}{(\Delta E/E)_{max}}}, \tag{47}$$

where

$$\left(\frac{\Delta E}{E}\right)^{n+1} = max_i \left(\frac{|E_i^{n+1} - E_i^n|}{E_i^{n+1} + E_0}\right), \tag{48}$$

where the parameter E_0 is an estimate for the lower bound of the energy density. Comparing Equation (47) to (45) we observed that Equation (45) is computationally more efficient. Therefore, we use Equation (45) in our numerical test problems.

4. Computational results

4.1 Smooth problem test
We use the LERH model to produce numerical results for this test problem. In this test, we run the code until a particular final time so that the computational solutions are free of shock waves and steep thermal fronts. The problem is to follow the evolution of the nonlinear waves that results from an initial energy deposition in a narrow region. The initial total energy density is given by

$$E(r,0) = \frac{\varepsilon_0 \exp(-r^2/c_0^2)}{(c_0 \sqrt{\pi})^3}, \tag{49}$$

where c_0 is a constant and set to $1/4$ for this test. Note that $c_0 \to 0$ gives a delta function at origin. We use the cell averaged values of E as in (Bates et al., 2001), i.e., we integrate (49) over the ith cell from $r_{i-1/2}$ to $r_{i+1/2}$ so that

$$E_i = \frac{\varepsilon_0[erf(r_{i+1/2}/c_0) - erf(r_{i-1/2}/c_0)] - 2\pi c_0^2[r_{i+1/2}E(r_{i+1/2}) - r_{i-1/2}E(r_{i-1/2})]}{\Delta V_i}, \tag{50}$$

where the symbol erf denotes the error function. The initial density is set to $\rho = 1/r$. The initial temperature is calculated by using $E = c_v \rho T + \frac{1}{2}\rho u$ where $u = 0$ initially. The boundary

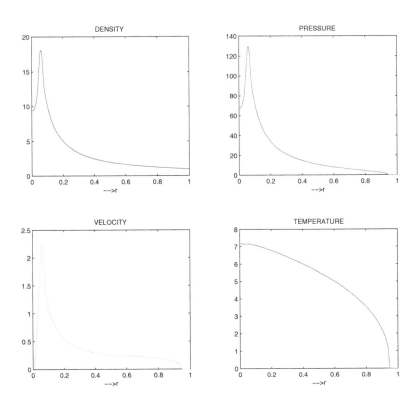

Fig. 3. Solution profiles resulting from the smooth problem test. The solutions are calculated for $t_{final} = 0.01$ with $M = 200$ cell points.

conditions for the hydrodynamics variables are *reflective* and *outflow* boundary conditions at the left and right ends of the computational domain respectively. The zero-flux boundary conditions are used for the temperature at both ends (e.g, $\partial T/\partial r|_{r=0} = 0$). The coefficient of thermal conduction is set to $\kappa(T) = T^{5/2}$.

We run the code until $t = 0.01$ with $\varepsilon_0 = 100$ using 200 cell points. The size of the computational domain is set to 1 (e.g, $R_0 = 1$ in Fig. 2). Fig. 3 shows the computed solutions for density, pressure, velocity, and temperature. As can be seen, there is no shock formation or steep thermal fronts occurred around this time. Fig. 4 shows our numerical time convergence analysis. To measure the rate of time convergence, we run the code with a fixed mesh (e.g, $M = 200$ cell points) and different time step refinements to a final time (e.g, $t = 0.01$). This provides a sequence of solution data ($E^{\Delta t}, E^{\Delta t/2}, E^{\Delta t/4}, \cdots$). Then we measure the L_2 norm of errors between *two* consecutive time step solutions ($\|E^{\Delta t} - E^{\Delta t/2}\|_2, \|E^{\Delta t/2} - E^{\Delta t/4}\|_2, \cdots$) and plot these errors against to a second order line. It is clear from Fig. 4 that we achieve second order time convergence unlike (Bates et al., 2001) fails to provide second order accurate results for the same test.

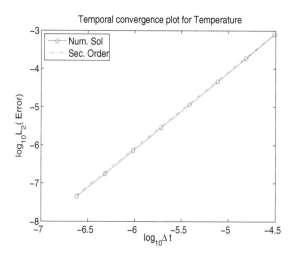

Fig. 4. Temporal convergence plot for the smooth problem test. $t_{final} = 0.01$ with $M = 200$ cell points.

4.2 Point explosion test

We use the HERH model for this test. We note that we have studied this test by using both of the LERH and HERH models and reported our results in two consecutive papers (Kadioglu & Knoll, 2010; Kadioglu, Knoll, Lowrie & Rauenzahn, 2010). This section reviews our numerical findings from (Kadioglu, Knoll, Lowrie & Rauenzahn, 2010). In this test, important physics such as the propagation of sharp shock discontinuities and steep thermal fronts occur. This is important, because this test enables us to study/determine the time accuracy of the strong numerical coupling of two distinct physical processes.

Typically a point explosion is characterized by the release of large amount of energy in a small region of space (few cells near the origin). Depending on the magnitude of the energy deposition, weak or strong explosions take place. If the initial explosion energy is not large enough, the diffusive effect is limited to region behind the shock. However, if the explosion energy is large, then the thermal front can precede the hydrodynamics front. Both weak and strong explosions are studied in (Kadioglu & Knoll, 2010) where the LERH model is considered. Here, we solve/recast the strong explosion test by using the HERH model. The problem setting is as follows. The initial total energy density is given by

$$E_0 = \frac{\varepsilon_0 \exp\left(-r^2/c_0^2\right)}{(c_0\sqrt{\pi})^3}, \tag{51}$$

where $\varepsilon_0 = 235$ and $c_0 = 1/300$. The initial fluid and radiation energies are set to $E(r,0) = E_v(r,0) = E_0/2$. The fluid density is initialized by $\rho(r,0) = r^{-19/9}$. The initial temperature is calculated by using $E = c_v\rho T/\gamma + \frac{1}{2}\rho u^2$ with the initial $u = 0$. The radiation diffusivity (κ in Equation (19)) is calculated by considering the LERH model and comparing it with the sum of Equation (18) plus $\mathcal{P}$ times Equation (19). For instance

$$\frac{\partial}{\partial t}(E + \mathcal{P}E_v) + \frac{1}{r^2}\frac{\partial}{\partial r}[r^2 u(E + p + \mathcal{P}(E_v + p_v))] = \frac{1}{r^2}\frac{\partial}{\partial r}(r^2\mathcal{P}\kappa\frac{\partial E_v}{\partial r}), \tag{52}$$

s compared to Equation (6) of (Kadioglu & Knoll, 2010). Then κ becomes

$$\kappa(\rho, T) = \kappa_0 \frac{\rho^a T^b}{4\mathcal{P}T^3},$$ (53)

where $\kappa_0 = 10^2$, $a = -2$ and $b = 13/2$ as in (Kadioglu & Knoll, 2010). We set $\mathcal{P} = 10^{-4}$ and $r_a = 10^8$ that appear in Equations (18) and (19). We compute the solutions until $t = 0.02$ using 400 cell points. Fig. 5 shows fluid density, fluid pressure, flow velocity, fluid energy, fluid temperature, and radiation temperature profiles. At this time ($t = 0.02$), hydrodynamical shocks are depicted near $r = 0.2$. In this test case, the thermal front (located near $r = 0.8$) propagates faster than the hydrodynamical shocks due to large initial energy deposition. Fig. 6 shows the time convergence analysis for different field variables. Clearly, we have obtained second order time accuracy for all variables. This convergence result is important, because this problem is a difficult one meaning that the coupling of different physics is highly non-linear and it is a challenge to produce fully second order convergence from an operator split method for these kinds of problems. One comment that can be made about our spatial discretization (LLF method), though it is not the primary focus of this study, is that our numerical results (figures in Fig. 5) indicate that the LLF fluxing procedure provides very good shock capturing with no spurious oscillations at or near the discontinuities.

4.3 Radiative shock test

The problem settings for this test are similar to (Drake, 2007; Lowrie & Edwards, 2008) where more precise physical definitions can be found. Radiative shocks are basically strong shock waves that the radiative energy flux plays essential role in the governing dynamics. Radiative shocks occur in many astrophysical systems where they move into an upstream medium leaving behind an altered downstream medium. In this test, we assume that a simple planar radiative shock exists normal to the flow as it is illustrated in Fig. 7. The initial shock profiles are determined by considering the given values in Region-1 and finding the values in Region-2 of Fig. 7. To find the values in Region-2, we use the so-called Rankine-Hugoniot relations or jump conditions (LeVeque, 1998; Smoller, 1994; Thomas, 1999). A general formula for the radiation hydrodynamics jump conditions is given in (Lowrie & Edwards, 2008). For instance

$$s(\rho_2 - \rho_1) = \rho_2 u_2 - \rho_1 u_1,$$ (54)

$$s(\rho_2 u_2 - \rho_1 u_1) = (\rho_2 u_2^2 + p_2 + \mathcal{P}p_{v,2}) - (\rho_1 u_1^2 + p_1 + \mathcal{P}p_{v,1}),$$ (55)

$$s(E_2 - E_1) = u_2(E_2 + p_2 + \mathcal{P}p_{v,2}) - u_1(E_1 + p_1 + \mathcal{P}p_{v,1}),$$ (56)

$$s(E_{v,2} - E_{v,1}) = u_2(E_{v,2}) - u_1(E_{v,1}),$$ (57)

where s is the propagation speed of the shock front. In our test problem, we assume that the radiation temperature is smooth. Therefore, it is sufficient to use the jump conditions for the compressible Euler equations to initiate hydrodynamics shock profiles. The Euler jump conditions can be easily obtained by dropping the radiative terms in Equations (54), (55), (56), and (57). Then the necessary formulae to initialize the shock solutions are

$$s = u_1 + c_1 \sqrt{1 + \frac{\gamma + 1}{2\gamma}(\frac{p_2}{p_1} - 1)},$$ (58)

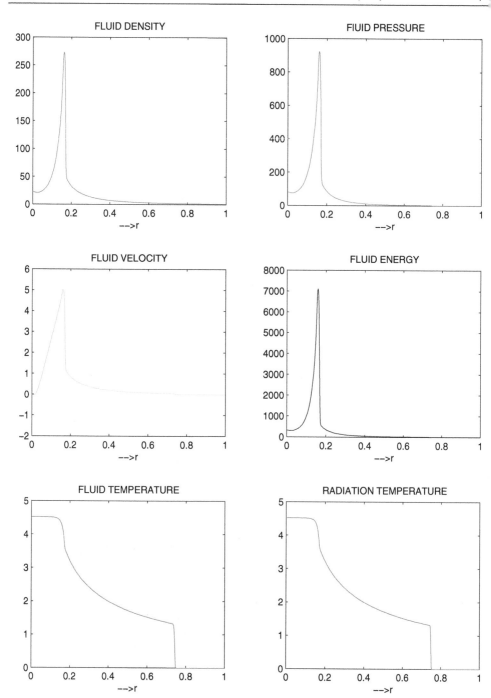

Fig. 5. Point explosion test with $t = 0.02$ and $M = 400$ cell points.

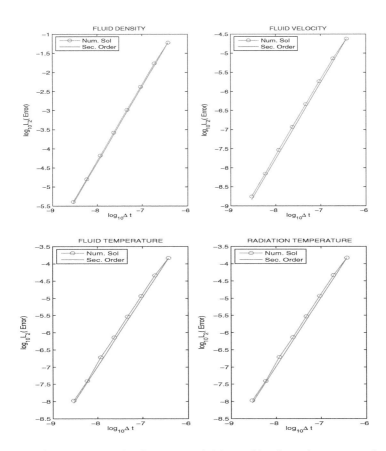

Fig. 6. Temporal convergence plot for various field variables from the point explosion test. = 0.001 and 400 cell points are used.

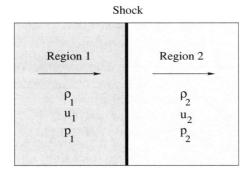

Fig. 7. A schematic diagram of a shock wave situation with the indicated density, velocity, and pressure for each region.

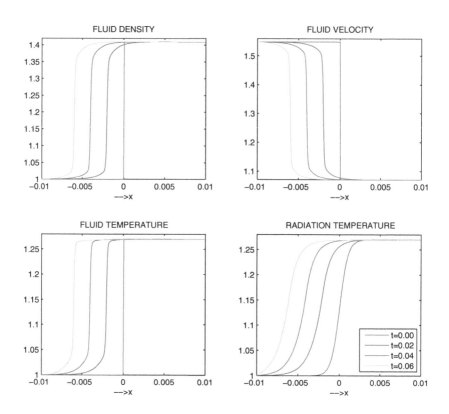

Fig. 8. Time evolution of the left moving radiative shock.

$$\frac{p_2}{p_1} = 1 + \frac{2\gamma}{\gamma+1}[(\frac{s-u_1}{c_1})^2 - 1], \tag{59}$$

$$u_2 = u_1 + \frac{p_2 - p_1}{\rho_1(s-u_1)}, \tag{60}$$

$$\frac{\rho_2}{\rho_1} = \frac{s-u_1}{s-u_2}, \tag{61}$$

where $c_1 = \sqrt{\gamma\frac{p_1}{\rho_1}}$ is the speed of sound in the fluid at upstream conditions. More details regarding the derivation of Equations (58)-(61) can be found in (Anderson, 1990; LeVeque, 1998; Smoller, 1994; Thomas, 1999; Wesseling, 2000).

We are interested in solving a left moving radiative shock problem. To achieve this, we set the initial shock speed $s = -0.1$ in Equation (58). Other upstream flow variables are set as follows; $\rho_1 = 1.0$, $T_1 = 1.0$, and $M_1 = u_1/c_1 = 1.2$ as the upstream Mach number. Then we calculate the pressure from a calorically perfect gas relation ($p_1 = R\rho_1 T_1$). Using p_1 and ρ_1, we calculate the upstream sound speed $c_1 = \sqrt{\gamma p_1/\rho_1}$ together with $u_1 = M_1 c_1$. With these information in hand, we can easily calculate the downstream values using Equations (58)-(61). The total fluid energies in both upstream and downstream directions are calculated

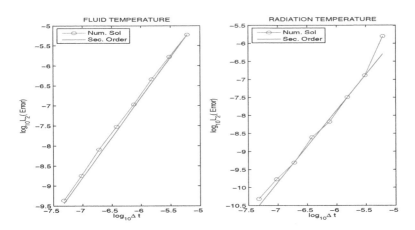

Fig. 9. Temporal convergence plot for the material and radiation temperature from the radiative shock test. $t = 0.02$ and 400 cell points are used.

by using the energy relation $E = c_v \rho T + \frac{1}{2} \rho u^2$. The radiation temperature is assumed to be a smooth function across the shock and equal to T_1 and T_2 on the left and right boundary of the computational domain, i.e., we choose

$$T_v(x,0) = \frac{(T_2 - T_1)}{2} tanh(1000x) + \frac{(T_2 + T_1)}{2}. \qquad (62)$$

The initial radiation energy is calculated by $E_v = T_v^4$. Other parameters that appear in Equations (16)-(19) are set as $\mathcal{P} = 10^{-4}$, $\sigma_a = 10^6$, and $\kappa = 1$. These parameters are chosen to be consistent with (Lowrie & Edwards, 2008). We solve Equations (16)-(19), the HERH model, in Cartesian coordinates with the above initial conditions. The solutions use fixed boundary conditions at both ends. In other words, at each time step, the solutions are reset to the initial boundary values. The numerical calculations are carried out with 400 cell points and $\Delta t = 10^{-6}$. Fig. 8 shows the time history of the solutions. Notice that the solutions are highly transient, therefore it is a good test to carry out a time convergence study. Fig. 9 shows time convergence analysis for the fluid and the radiation temperature. Second order time convergence can be clearly seen in both fields.

5. Convergence analysis

In this section, we present a mathematical analysis (modified equation analysis) to study the analytical convergence behavior of our self-consistent IMEX method and compare it to a classic IMEX method. The modified equation analysis (truncation error analysis) is performed by considering the LERH model (Equations (8)-(10) or (20)-(21)). Also, for simplicity, we assume that the system given by Equations (20)-(21) is written in cartesian coordinates. In the introduction, we first described a classic IMEX approach then presented our self-consistent IMEX method. Therefore, we shall follow the same order in regards to below mathematical analysis.

5.1 A classic IMEX method

The classic IMEX method operates on Equations (20)-(21) as follows.
Explicit block:
Step-1:

$$\rho^1 = \rho^n - \Delta t \frac{\partial}{\partial x}(\rho u)^n,$$

$$(\rho u)^1 = (\rho u)^n - \Delta t \frac{\partial}{\partial x}[(\rho u^2)^n + p^n],$$

$$E^1 = E^n - \Delta t \frac{\partial}{\partial x}\{u^n[E^n + p^n]\}, \tag{63}$$

Step-2:

$$\rho^{n+1} = \frac{\rho^1 + \rho^n}{2} - \frac{\Delta t}{2}\frac{\partial}{\partial x}(\rho u)^1,$$

$$(\rho u)^{n+1} = \frac{(\rho u)^1 + (\rho u)^n}{2} - \frac{\Delta t}{2}\frac{\partial}{\partial x}[\rho^1(u^2)^1 + R\rho^1\mathbf{T^n}],$$

$$E^* = \frac{E^n + E^1}{2} - \frac{\Delta t}{2}\frac{\partial}{\partial x}\{u^1[c_v\rho^1\mathbf{T^n} + \frac{1}{2}\rho^1(u^1)^2 + R\rho^1\mathbf{T^n}]\}, \tag{64}$$

Implicit block:

$$\frac{E^{n+1} - E^*}{\Delta t} = \frac{1}{2}\frac{\partial}{\partial x}(\kappa^{n+1}\frac{\partial T^{n+1}}{\partial x}) + \frac{1}{2}\frac{\partial}{\partial x}(\kappa^n\frac{\partial T^n}{\partial x}), \tag{65}$$

where we incorporated with the equation of states relations plus we assume that the explicit block is based on a second order TVD Runga-Kutta method and the implicit block is similar to the Crank-Nicolson method. Notice that the classic IMEX method is executed in such a way that the implicit temperature does not impact the explicit block (refer to the highlighted terms in Equation (64)). We carry out the modified equation analysis for the energy part of Equations (63)-(65), but the same procedure can easily be extended to the whole system. We consider

$$E^1 = E^n - \Delta t \frac{\partial}{\partial x}\{u^n[E^n + p^n]\}, \tag{66}$$

$$E^* = \frac{E^n + E^1}{2} - \frac{\Delta t}{2}\frac{\partial}{\partial x}\{u^1[c_v\rho^1 T^n + \frac{1}{2}\rho^1(u^1)^2 + R\rho^1 T^n]\}, \tag{67}$$

Substituting Equation (66) into (67), we get

$$E^* = E^n - \frac{\Delta t}{2}\frac{\partial}{\partial x}\{u^n[E^n + p^n]\} - \frac{\Delta t}{2}\frac{\partial}{\partial x}\{u^1[c_v\rho^1 T^n + \frac{1}{2}\rho^1(u^1)^2 + R\rho^1 T^n]\}. \tag{68}$$

We let $L(E^n) = -\frac{\partial}{\partial x}\{u^n[E^n + p^n]\}$ and use $T^n = T^1 - \Delta t T_t^n + O(\Delta t^2)$, then (68) becomes

$$E^* = E^n + \frac{\Delta t}{2}L(E^n) - \frac{\Delta t}{2}\frac{\partial}{\partial x}\{u^1[c_v\rho^1(T^1 - \Delta t\frac{\partial T^n}{\partial t} + O(\Delta t^2))$$

$$+ \frac{1}{2}\rho^1(u^1)^2 + R\rho^1(T^1 - \Delta t\frac{\partial T^n}{\partial t} + O(\Delta t^2))]\}. \tag{69}$$

n IMEX Method for the Euler Equations That Posses Strong Non-Linear Heat Conduction and Stiff Source Terms
Radiation Hydrodynamics)

23

Carrying out the necessary algebra, Equation (69) becomes

$$E^* = E^n + \frac{\Delta t}{2}L(E^n) + \frac{\Delta t}{2}L(E^1) + \frac{\Delta t^2}{2}\rho^1 u^1 \frac{\partial T^n}{\partial t}(c_v + R) + O(\Delta t^3), \tag{70}$$

where $L(E^1) = -\frac{\partial}{\partial x}\{u^1[c_v\rho^1 T^1 + \frac{1}{2}\rho^1(u^1)^2 + R\rho^1 T^1]\} = -\frac{\partial}{\partial x}\{u^1[E^1 + p^1]\}$. Further simplification comes from the following identity

$$L(E^1) = L(E^n) + \Delta t\frac{\partial L}{\partial t} + O(\Delta t^2). \tag{71}$$

Inserting Equation (71) into (70), we get

$$E^* = E^n + \Delta t L(E^n) + \frac{\Delta t^2}{2}\frac{\partial L}{\partial t} + \frac{\Delta t^2}{2}\rho^1 u^1 \frac{\partial T^n}{\partial t}(c_v + R) + O(\Delta t^3). \tag{72}$$

Now, we consider the following Taylor series for the implicit block (Equation (65))

$$E^{n+1} = E^n + \Delta t\frac{\partial E^n}{\partial t} + \frac{\Delta t^2}{2}\frac{\partial^2 E^n}{\partial t^2} + O(\Delta t^3), \tag{73}$$

$$T^{n+1} = T^n + \Delta t\frac{\partial T^n}{\partial t} + \frac{\Delta t^2}{2}\frac{\partial^2 T^n}{\partial t^2} + O(\Delta t^3), \tag{74}$$

$$\kappa^{n+1} = \kappa^n + \Delta t\frac{\partial \kappa^n}{\partial t} + \frac{\Delta t^2}{2}\frac{\partial^2 \kappa^n}{\partial t^2} + O(\Delta t^3). \tag{75}$$

Substituting Equations (73), (74), (75), and (72) into Equation (65), we form the truncation term as

$$\tau^n = E^n + \Delta t\frac{\partial E^n}{\partial t} + \frac{\Delta t^2}{2}\frac{\partial^2 E^n}{\partial t^2} - [E^n + \Delta t L(E^n) + \frac{\Delta t^2}{2}\frac{\partial L(E^n)}{\partial t}$$
$$+ \frac{\Delta t^2}{2}\rho^1 u^1 \frac{\partial T^n}{\partial t}(c_v + R)] - \Delta t\frac{1}{2}\frac{\partial}{\partial x}[(\kappa^n + \Delta t\frac{\partial \kappa^n}{\partial t} + \frac{\Delta t^2}{2}\frac{\partial^2 \kappa^n}{\partial t^2})$$
$$\frac{\partial}{\partial x}(T^n + \Delta t\frac{\partial T^n}{\partial t} + \frac{\Delta t^2}{2}\frac{\partial^2 T^n}{\partial t^2})] - \Delta t\frac{1}{2}\frac{\partial}{\partial x}(\kappa^n \frac{\partial T^n}{\partial x}) + O(\Delta t^3). \tag{76}$$

Cancelling the opposite sign common terms and grouping the other terms together, we get

$$\tau^n = \Delta t\,[\frac{\partial E^n}{\partial t} - L(E^n)] + \frac{\Delta t^2}{2}\frac{\partial}{\partial t}[\frac{\partial E^n}{\partial t} - L(E^n)]$$
$$- [\frac{\Delta t}{2}\frac{\partial}{\partial x}(\kappa^n \frac{\partial T^n}{\partial x}) + \frac{\Delta t}{2}\frac{\partial}{\partial x}(\kappa^n \frac{\partial T^n}{\partial x})]$$
$$- [\frac{\Delta t^2}{2}\frac{\partial}{\partial x}(\kappa^n \frac{\partial T_t^n}{\partial x}) + \frac{\Delta t^2}{2}\frac{\partial}{\partial x}(\kappa_t^n \frac{\partial T^n}{\partial x})]$$
$$+ \frac{\Delta t^2}{2}\rho^1 u^1 \frac{\partial T^n}{\partial t}(c_v + R) + O(\Delta t^3). \tag{77}$$

This further simplifies by using

$$\frac{\partial}{\partial t}[\frac{\partial}{\partial x}(\kappa^n \frac{\partial T^n}{\partial x})] = \frac{\partial}{\partial x}(\kappa_t^n \frac{\partial T^n}{\partial x}) + \frac{\partial}{\partial x}(\kappa^n \frac{\partial T_t^n}{\partial x}). \tag{78}$$

Then we have

$$\tau^n = \Delta t \left[\frac{\partial E^n}{\partial t} - L(E^n) - \frac{\partial}{\partial x}(\kappa^n \frac{\partial T^n}{\partial x}) \right]$$

$$+ \frac{\Delta t^2}{2} \frac{\partial}{\partial t} \left[\frac{\partial E^n}{\partial t} - L(E^n) - \frac{\partial}{\partial x}(\kappa^n \frac{\partial T^n}{\partial x}) \right]$$

$$+ \frac{\Delta t^2}{2} \rho^1 u^1 \frac{\partial T^n}{\partial t} (c_v + R) + O(\Delta t^3). \tag{79}$$

From the energy equation (Equation (10)) we have $\frac{\partial E^n}{\partial t} - L(E^n) - \frac{\partial}{\partial x}(\kappa^n \frac{\partial T^n}{\partial x}) = 0$, thus Equation (79) becomes

$$\tau^n = \frac{\Delta t^2}{2} \rho^1 u^1 \frac{\partial T^n}{\partial t}(c_v + R) + O(\Delta t^3). \tag{80}$$

This shows that the classic IMEX method carries first order terms in the resulting truncation error. This conclusion will be verified by our numerical computations (refer to Fig. 10).

5.2 A self-consistent IMEX method
We have already described how the self-consistent IMEX method operates on Equations (20)-(21) in Section 3.1. However, to be able to easily follow the analysis, we repeat the self-consistent operator splitting below.
Explicit block:
Step-1:

$$\rho^1 = \rho^n - \Delta t \frac{\partial}{\partial x}(\rho u)^n,$$

$$(\rho u)^1 = (\rho u)^n - \Delta t \frac{\partial}{\partial x}[(\rho u^2)^n + p^n],$$

$$E^1 = E^n - \Delta t \frac{\partial}{\partial x}\{u^n[E^n + p^n]\}, \tag{81}$$

Step-2:

$$\rho^{n+1} = \frac{\rho^1 + \rho^n}{2} - \frac{\Delta t}{2} \frac{\partial}{\partial x}(\rho u)^1,$$

$$(\rho u)^{n+1} = \frac{(\rho u)^1 + (\rho u)^n}{2} - \frac{\Delta t}{2} \frac{\partial}{\partial x}[\rho^1(u^2)^1 + R\rho^1 \mathbf{T^{n+1}}],$$

$$E^* = \frac{E^n + E^1}{2} - \frac{\Delta t}{2} \frac{\partial}{\partial x}\{u^1[c_v\rho^1 \mathbf{T^{n+1}} + \frac{1}{2}\rho^1(u^1)^2 + R\rho^1 \mathbf{T^{n+1}}]\}, \tag{82}$$

Implicit block:

$$\frac{E^{n+1} - E^*}{\Delta t} = \frac{1}{2} \frac{\partial}{\partial x}(\kappa^{n+1} \frac{\partial T^{n+1}}{\partial x}) + \frac{1}{2} \frac{\partial}{\partial x}(\kappa^n \frac{\partial T^n}{\partial x}). \tag{83}$$

Notice that the implicit temperature impacts the explicit block in this case (refer to the highlighted terms in Equation (82)). Again, we perform the modified equation analysis on the energy part of Equations (81)-(83). Substituting E^1 into E^*, we get

$$E^* = E^n - \frac{\Delta t}{2} \frac{\partial}{\partial x}\{u^n[E^n + p^n]\} - \frac{\Delta t}{2} \frac{\partial}{\partial x}\{u^1[c_v\rho^1 T^{n+1} + \frac{1}{2}\rho^1(u^1)^2 + R\rho^1 T^{n+1}]\}. \tag{84}$$

We let $L(E^n) = -\frac{\partial}{\partial x}\{u^n[E^n + p^n]\}$ and use Equation (74) in (84) to get

$$E^* = E^n + \frac{\Delta t}{2}L(E^n) - \frac{\Delta t}{2}\frac{\partial}{\partial x}\{u^1[c_v\rho^1(T^n + \Delta t\frac{\partial T^n}{\partial t} + O(\Delta t^2))$$

$$+ \frac{1}{2}\rho^1(u^1)^2 + R\rho^1(T^n + \Delta t\frac{\partial T^n}{\partial t} + O(\Delta t^2))]\}. \tag{85}$$

Now, we insert $T^n = T^1 - \Delta t\frac{\partial T^n}{\partial t} + O(\Delta t^2)$ in Equation (85) and perform few algebra to get

$$E^* = E^n + \frac{\Delta t}{2}L(E^n) + \frac{\Delta t}{2}L(E^1) + O(\Delta t^3), \tag{86}$$

where $L(E^1) = -\frac{\partial}{\partial x}\{u^1[c_v\rho^1 T^1 + \frac{1}{2}\rho^1(u^1)^2 + R\rho^1 T^1]\}$. Equation (86) can be further simplified by using (71),

$$E^* = E^n + \Delta t L(E^n) + \frac{\Delta t^2}{2}\frac{\partial L}{\partial t} + O(\Delta t^3). \tag{87}$$

Making use of the Taylor series given in Equations (73), (74), (75), and Equation (87) in the implicit discretization (83), we form the truncation error term as

$$\tau^n = E^n + \Delta t\frac{\partial E^n}{\partial t} + \frac{\Delta t^2}{2}\frac{\partial^2 E^n}{\partial t^2} - (E^n + \Delta t L(E^n) + \frac{\Delta t^2}{2}\frac{\partial L(E^n)}{\partial t})$$

$$- \Delta t\frac{1}{2}\frac{\partial}{\partial x}[(\kappa^n + \Delta t\frac{\partial \kappa^n}{\partial t} + \frac{\Delta t^2}{2}\frac{\partial^2 \kappa^n}{\partial t^2})\frac{\partial}{\partial x}(T^n + \Delta t\frac{\partial T^n}{\partial t} + \frac{\Delta t^2}{2}\frac{\partial^2 T^n}{\partial t^2})]$$

$$- \Delta t\frac{1}{2}\frac{\partial}{\partial x}(\kappa^n\frac{\partial T^n}{\partial x}) + O(\Delta t^3). \tag{88}$$

Cancelling the opposite sign common terms, grouping the other terms together, and making use of Equation (78), we get

$$\tau^n = \Delta t\,[\frac{\partial E^n}{\partial t} - L(E^n) - \frac{\partial}{\partial x}(\kappa^n\frac{\partial T^n}{\partial x})] + \frac{\Delta t^2}{2}\frac{\partial}{\partial t}[\frac{\partial E^n}{\partial t} - L(E^n) - \frac{\partial}{\partial x}(\kappa^n\frac{\partial T^n}{\partial x})] + O(\Delta t^3) \tag{89}$$

Again from the energy equation, we know that $\frac{\partial E^n}{\partial t} - L(E^n) - \frac{\partial}{\partial x}(\kappa^n\frac{\partial T^n}{\partial x}) = 0$, thus Equation (89) becomes

$$\tau^n = O(\Delta t^3), \tag{90}$$

clearly proving that the self-consistent IMEX method is second order.

Here, we numerically verify our analytical findings about the two IMEX approaches. We solve the point explosion problem studied in (Kadioglu & Knoll, 2010) by using the LERH model and $M = 200$ cell points until the final time $t = 0.02$. We note that we ran the code twice as longer final time than the original test in order to allow the numerical methods to depict more accurate time behaviors. In Fig. 10, we plot the L_2-norm of errors for variety of flow variables committed by the both approaches. Fig. 10 clearly shows that the classic IMEX method suffers from order reductions as predicted by our mathematical analysis. We present more detailed analysis regarding more general IMEX methods (e.g., Strang splitting type methods (Knoth & Wolke, 1999; Strang, 1968)) in our forthcoming paper (Kadioglu, Knoll & Lowrie, 2010).

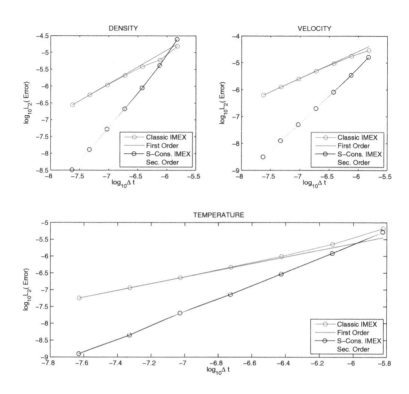

Fig. 10. The self-consistent IMEX method versus a classic IMEX method in terms of the time convergence.

6. Conclusion

We have presented a self-consistent implicit/explicit (IMEX) time integration technique for solving the Euler equations that posses strong nonlinear heat conduction and very stiff source terms (Radiation hydrodynamics). The key to successfully implement an implicit/explicit algorithm in a self-consistent sense is to carry out the explicit integrations as part of the non-linear function evaluations within the implicit solver. In this way, the improved time accuracy of the non-linear iterations is immediately felt by the explicit algorithm block and the more accurate explicit solutions are readily available to form the next set of non-linear residuals. We have solved several test problems that use both of the low and high energy density radiation hydrodynamics models (the LERH and HERH models) in order to validate the numerical order of accuracy of our scheme. For each test, we have established second order time convergence. We have also presented a mathematical analysis that reveals the analytical behavior of our method and compares it to a classic IMEX approach. Our analytical findings have been supported/verified by a set of computational results. Currently, we are exploring more about our multi-phase IMEX study to solve multi-phase flow systems that posses tight non-linear coupling between the interface and fluid dynamics.

Acknowledgement

he submitted manuscript has been authored by a contractor of the U.S. Government under ontract No. DEAC07-05ID14517 (INL/MIS-11-22498). Accordingly, the U.S. Government tains a non-exclusive, royalty-free license to publish or reproduce the published form of this ontribution, or allow others to do so, for U.S. Government purposes.

References

nderson, J. (1990). *Modern Compressible Flow.*, Mc Graw Hill.

scher, U. M., Ruuth, S. J. & Spiteri, R. J. (1997). Implicit-explicit Runge-Kutta methods for time-dependent partial differential equations., *Appl. Num. Math.* 25: 151–167.

scher, U. M., Ruuth, S. J. & Wetton, B. T. R. (1995). Implicit-explicit methods for time-dependent pde's., *SIAM J. Num. Anal.* 32: 797–823.

ates, J. W., Knoll, D. A., Rider, W. J., Lowrie, R. B. & Mousseau, V. A. (2001). On consistent time-integration methods for radiation hydrodynamics in the equilibrium diffusion limit: Low energy-density regime., *J. Comput. Phys.* 167: 99–130.

owers, R. & Wilson, J. (1991). *Numerical Modeling in Applied Physics and Astrophysics.*, Jones & Bartlett, Boston.

rown, P. & Saad, Y. (1990). Hybrid Krylov methods for nonlinear systems of equations., *SIAM J. Sci. Stat. Comput.* 11: 450–481.

astor, J. (2006). *Radiation Hydrodynamics.*, Cambridge University Press.

)ai, W. & Woodward, P. (1998). Numerical simulations for radiation hydrodynamics. i. diffusion limit., *J. Comput. Phys.* 142: 182.

)embo, R. (1982). Inexact newton methods., *SIAM J. Num. Anal.* 19: 400–408.

)rake, R. (2007). Theory of radiative shocks in optically thick media., *Physics of Plasmas* 14: 43301.

nsman, L. (1994). Test problems for radiation and radiation hydrodynamics codes., *The Astrophysical Journal* 424: 275–291.

;ottlieb, S. & Shu, C. (1998). Total variation diminishing Runge-Kutta schemes., *Mathematics of Computation* 221: 73–85.

;ottlieb, S., Shu, C. & Tadmor, E. (2001). Strong stability-preserving high-order time discretization methods., *Siam Review* 43-1: 89–112.

.adioglu, S. & Knoll, D. (2010). A fully second order Implicit/Explicit time integration technique for hydrodynamics plus nonlinear heat conduction problems., *J. Comput. Phys.* 229-9: 3237–3249.

.adioglu, S. & Knoll, D. (2011). Solving the incompressible Navier-Stokes equations by a second order IMEX method with a simple and effective preconditioning strategy., *Appl. Math. Comput. Sci.* Accepted.

.adioglu, S., Knoll, D. & Lowrie, R. (2010). Analysis of a self-consistent IMEX method for tightly coupled non-linear systems., *SIAM J. Sci. Comput.* to be submitted.

.adioglu, S., Knoll, D., Lowrie, R. & Rauenzahn, R. (2010). A second order self-consistent IMEX method for Radiation Hydrodynamics., *J. Comput. Phys.* 229-22: 8313–8332.

.adioglu, S., Knoll, D. & Oliveria, C. (2009). Multi-physics analysis of spherical fast burst reactors., *Nuclear Science and Engineering* 163: 1–12.

.adioglu, S., Knoll, D., Sussman, M. & Martineau, R. (2010). A second order JFNK-based IMEX method for single and multi-phase flows., *Computational Fluid Dynamics, Springer-Verlag* DOI 10.1007/978 − 3 − 642 − 17884 − 9_69.

Kadioglu, S., Sussman, M., Osher, S., Wright, J. & Kang, M. (2005).   A Second Order Primitive Preconditioner For Solving All Speed Multi-Phase Flows., *J. Comput. Phys* 209-2: 477–503.

Kelley, C. (2003). *Solving Nonlinear Equations with Newton's Method.*, Siam.

Khan, L. & Liu, P. (1994).   An operator-splitting algorithm for coupled one-dimensional advection-diffusion-reaction equations., *Computer Meth. Appl. Mech. Eng* 127: 181–201.

Kim, J. & Moin, P. (1985).   Application of a fractional-step method to incompressible navier-stokes equations., *J. Comput. Phys.* 59: 308–323.

Knoll, D. A. & Keyes, D. E. (2004).   Jacobian-free Newton Krylov methods: a survey of approaches and applications., *J. Comput. Phys.* 193: 357–397.

Knoth, O. & Wolke, R. (1999).   Strang splitting versus implicit-explicit methods in solving chemistry transport models., *Transactions on Ecology and the Environment* 28: 524–528.

LeVeque, R. (1998). *Finite Volume Methods for Hyperbolic Problems.*, Cambridge University Press , Texts in Applied Mathematics.

Lowrie, R. B., Morel, J. E. & Hittinger, J. A. (1999).   The coupling of radiation and hydrodynamics., *Astrophys. J.* 521: 432.

Lowrie, R. & Edwards, J. (2008).   Radiative shock solutions with grey nonequilibrium diffusion., *Shock Waves* 18: 129–143.

Lowrie, R. & Rauenzahn, R. (2007).   Radiative shock solutions in the equilibrium diffusion limit., *Shock Waves* 16: 445–453.

Marshak, R. (1958). Effect of radiation on shock wave behavior., *Phys. Fluids* 1: 24–29.

Mihalas, D. & Mihalas, B. (1984). *Foundations of Radiation Hydrodynamics.*, Oxford University Press, New York.

Pomraning, G. (1973). *The Equations of Radiation Hydrodynamics.*, Pergamon, Oxford.

Reid, J. (1971). *On the methods of conjugate gradients for the solution of large sparse systems of linear equations.*, Large Sparse Sets of Linear Equations, Academic Press, New York.

Rider, W. & Knoll, D. (1999).   Time step size selection for radiation diffusion calculations., *J Comput. Phys.* 152-2: 790–795.

Ruuth, S. J. (1995).   Implicit-explicit methods for reaction-diffusion problems in pattern formation., *J. Math. Biol.* 34: 148–176.

Saad, Y. (2003). *Iterative Methods for Sparse Linear Systems.*, Siam.

Shu, C. & Osher, S. (1988).   Efficient implementation of essentially non-oscillatory shock capturing schemes., *J. Comput. Phys.* 77: 439.

Shu, C. & Osher, S. (1989).   Efficient implementation of essentially non-oscillatory shock capturing schemes II., *J. Comput. Phys.* 83: 32.

Smoller, J. (1994). *Shock Waves And Reaction-diffusion Equations.*, Springer.

Strang, G. (1968). On the construction and comparison of difference schemes., *SIAM J. Numer. Anal.* 8: 506–517.

Strikwerda, J. (1989). *Finite Difference Schemes Partial Differential Equations.*, Wadsworth & Brooks/Cole, Advance Books & Software, Pacific Grove, California.

Thomas, J. (1998). *Numerical Partial Differential Equations I (Finite Difference Methods).*, Springer-Verlag New York, Texts in Applied Mathematics.

Thomas, J. (1999). *Numerical Partial Differential Equations II (Conservation Laws and Elliptic Equations).*, Springer-Verlag New York, Texts in Applied Mathematics.

Wesseling, P. (2000).   *Principles of Computational Fluid Dynamics.*, Springer Series in Computational Mathematics.

2

Electro-Hydrodynamics of Micro-Discharges in Gases at Atmospheric Pressure

O. Eichwald, M. Yousfi, O. Ducasse, N. Merbahi,
J. P. Sarrette, M. Meziane and M. Benhenni
University of Toulouse, University Paul Sabatier,
LAPLACE Laboratory,
France

1. Introduction

Micro-discharges are specific cold filamentary plasma that are generated at atmospheric pressure between electrodes stressed by high voltage. As cold plasma or non-thermal plasma, we suggest that the energy of electrons inside the conductive plasma is much higher than the energy of the heaviest particles (molecules and ions). In such kind of plasma, the temperature of the gas remains cold (i.e. more or less equal to the ambient temperature) unlike in the field of thermal plasmas where the gas temperature can reach some thousands of Kelvin. This high level of temperature can be measured for example in plasma torch or in lightning.

The conductive channels of micro-discharges are very thin. Their diameters are estimated around some tens of micrometers. This specificity explains their name: micro-discharge. Another of their characteristic is their very fast development. In fact, micro-discharges propagate at velocity that can attain some tens of millimetres per nanosecond i.e. some 10^7 cm.s^{-1}. This very fast velocity is due to the propagation of space charge dominated streamer heads. The space charge inside the streamer head creates a very high electric field in which the electrons are accelerated like in an electron gun. These electrons interact with the gas and create mainly ions and radicals. In fact, the energy distribution of electrons inside streamer heads favours the chemical electron-molecule reactions rather than the elastic electron-molecule collisions. Therefore, micro-discharges are mainly used in order to activate chemical reactions either in the gas volume or on a surface (Penetrante & Schultheis, 1993, Urashima &Chang, 2010, Foest et al. 2005, Clement, 2001).

Several designs of plasma reactors are able to generate micro-discharges. The most convenient and the well known is probably the corona discharge reactor (Loeb, 1961&1965, Winands, 2006, Ono & Oda a, 2004, van Veldhuizen & Rutgers, 2002, Briels et al., 2006). Corona micro-discharges reactor has at least two asymmetric electrodes i.e. with one of them presenting a low curvature that introduces a pin effect where the geometric electric field is enhanced. The corona micro-discharges are initiated from this high geometric field area. Some samples of corona reactor geometries are shown in Fig. 1.

The transient character and the small dimensions make some micro-discharges parameters, like charged and radical densities, electron energy or electric field strength, difficult to be accessible to measurements. Therefore, the complete simulation of the discharge reactor, in complement to experimental study can lead to a better understanding of the physico-

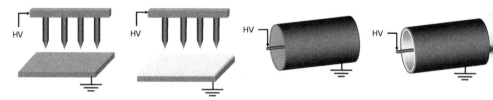

Fig. 1. Sample of pin-to-plane and wire-to-cylinder corona discharge reactors. The light blue material corresponds to a dielectric material. Depending on applications, design and reactor efficiency, the High Voltage (HV) shape can be DC, pulsed, AC or a combination of them.

chemical activity triggered by the micro-discharge during the plasma process. All these information can be used in order to improve the reactor design and to achieve the best operating conditions (such as the reactor geometry, the flue gas resident time, the applied voltage shape and magnitude, among others) as a function of the chosen applications.

The present chapter is devoted to description of the main electro-hydrodynamics phenomena that take place in non-thermal plasma reactors at atmospheric pressure activated by corona micro-discharges. The first section describes the micro-discharges characteristics using the experimental results obtained in a mono pin-to-plane reactor stressed by either DC or pulsed high voltage. The physics of the micro-discharges development is explained and a complete hydrodynamics model is proposed based on the moments of Boltzmann equations for charged and neutral particles. Then before to conclude, the previous described model is used in order to simulate the strongly coupled chemical and hydrodynamics phenomena generated by micro-discharges in a non thermal plasma reactor.

2. Description of positive corona micro-discharges

2.1 Introduction
In this first section, we describe the main characteristics of the corona micro-discharge formation and development as a function of several operating parameters such as the geometry of electrodes or the shape and magnitude of applied high voltage. Then, based on Boltzmann kinetic theory, we describe the strongly coupled electrical, hydrodynamics and chemical phenomena that take place in a compressible gas crossed by micro-discharges.

2.2 Positive corona micro-discharge under DC voltage condition
Let consider a mono pin-to-plane electrode corona reactor filled with dry air at atmospheric pressure and ambient temperature (Dubois et al., 2007). A DC high voltage supply is connected to the pin through a mega ohm resistor. When the applied voltage is raised gradually there is no sustained discharge current as much as the electrical gap field remains less than the onset one. Then, a sudden current pulse appears marking the beginning of the self sustained onset streamer regime. The associated current pulses occur intermittently and randomly and the mean current is very low (of few μA). Using a CCD camera with a large time shutter, we can observe a low intensity spot light just around the pin (see Fig. 2a). If we continue to increase the DC voltage, the current pulses vanish. However, the spot light near the point is always observed but with a quite higher intensity (see Fig. 2b). This regime corresponds to the classical glow corona discharge which is characterised by a drift of charged particles in the inter-electrode gap. The average current can reach some tens of μA.

For a high voltage threshold value, some regular repetitive current pulses appear with a repetition frequency of some tens of kHz and a magnitude of some tens to hundred of mA. Each current pulse lasts some hundred of nanoseconds and corresponds to the propagation of a mono-filament corona micro-discharge shown in Fig. 2c.

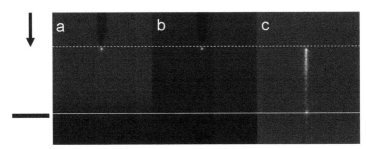

Fig. 2. Photography of the different corona discharge regimes under positive DC voltage condition (inter-electrode distance = 7mm, pin radius = 20 µm, dry air, atmospheric pressure). a: onset streamer, DC voltage magnitude = 3.2kV, time camera shutter = 1s, b: glow discharge, DC voltage magnitude = 5kV, time camera shutter = 10ms, c: streamer micro-discharge, DC voltage magnitude = 7.2kV, time camera shutter = 10µs (Eichwald et al., 2008).

More detailed information on the spatio-temporal evolution of the micro-discharge can be obtained thanks to the analysis of the streak camera picture shown in Fig. 3 and the corresponding current pulse shown in Fig. 4 (Eichwald et al., 2008, Marode, 1975). In Fig. 3, the X-axis is the time axis while the Y-axis is the inter-electrode distance. The electrode location is shown in the drawing at the left side of Fig. 3. For a given time on the X-axis, the light emission of the micro-discharge filament at each position is focused along the corresponding Y-axis coordinate. When 8.2kV is applied to the pin, three main phases can be distinguished in the corona micro-discharge development. The first one corresponds to the primary streamer propagation from the anode pin towards the cathode plane. The primary streamer propagates a luminous spot (called streamer head) which leaves the first narrow luminous trail shown on the streak picture of Fig. 3. During this first phase, the current rapidly increases as shown in Fig. 4 between 50ns and 75ns. The second phase

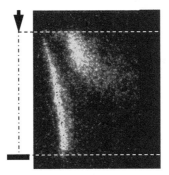

Fig. 3. Streak camera picture of a corona micro-discharge: Inter-electrode distance = 7mm, pin radius = 20 µm, dry air, atmospheric pressure, DC voltage magnitude = 8.2kV

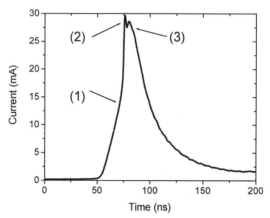

Fig. 4. Instantaneous micro-discharge current (inter-electrode distance = 7mm, pin radius = 20 μm, dry air, atmospheric pressure, DC voltage magnitude = 8.2kV)

corresponds to the arrival of the primary streamer at the cathode. It is associated to both the sudden increase of the current pulse at around 75ns (see Fig. 4) and the first current peak. In the present experimental conditions, the current pulse magnitude reaches a maximum of 30mA. On the current curve of Fig. 4, we also observe that the primary streamer needs about 25ns to cross the inter electrode gap and to reach the cathode plane 7mm underneath the pin. Thereby, the mean primary streamer velocity can be estimated of about 3×10^7cm s^{-1}. We also observe in Fig. 3 the development of a secondary streamer (Sigmond, 1984) starting from the point when the primary streamer arrives at the cathode plane. The associated light emission is more diffuse on the streak picture because the radiative species are distributed along the pre-ionized channel. The development and propagation of the secondary streamer induce a second current peak (see Fig. 4). Finally, each current pulse is characterised by the propagation of primary and secondary streamers which in turn create the thin ionized channels of the micro-discharge shown in Fig. 2c.

2.3 Positive corona micro-discharge under pulsed voltage condition

The morphology of micro-discharges under pulse voltage condition is quite different from the case of DC voltage condition (van Veldhuizen & Rutgers, 2002, Abahazem et al. 2008). However, we will see at the end of the present section the correspondence between both regimes using a large voltage pulse width. Fig. 5 shows a sample of a high voltage pulse applied on the pin of a pin-to-plane corona reactor and the resulting measured current pulse. The pulse voltage width is first chosen in order to obtain only one micro-discharge per pulse. The experimental conditions are very similar to those used for the DC voltage study described in previous section 2.2. In Fig. 5, the two current peaks superposed with the increasing and decreasing fronts of the pulse voltage are two capacitive current pulses generated by the equivalent capacitance of the pin-to-plane electrode configuration. The micro-discharge current pulse is positioned at time t=0ns in Fig. 5. A detailed description of this peculiar current pulse can be seen in Fig. 6. A rapid comparison with the DC current pulse in Fig. 4 indicates that the current pulse magnitude under pulse voltage condition is much higher (~175mA) than in the DC voltage case (~30mA). In fact, the ICCD time

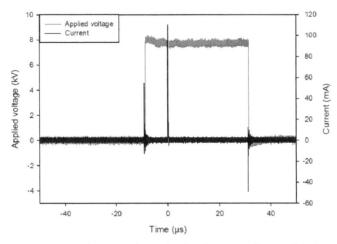

Fig. 5. Instantaneous measured current for pulsed voltage conditions: Maximum voltage magnitude=8kV, pulse voltage width=40μs, inter-electrode distance=8mm, pin radius=25μm, dry air at atmospheric pressure.

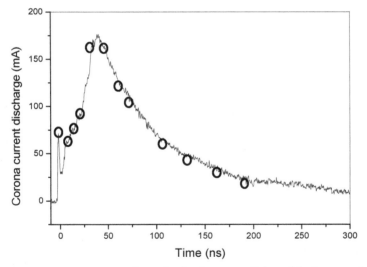

Fig. 6. Instantaneous corona current for pulsed voltage conditions: Maximum voltage magnitude=8kV, pulse voltage width=40μs, inter-electrode distance=8mm, pin radius=25μm, dry air at atmospheric pressure.

integrated picture of Fig. 7 clearly shows that there are several streamers starting from the pin towards the plane. This branching mechanism occurs in pulsed voltage conditions and therefore gives a higher discharge current than in the case of DC voltage. The evolution of the corona current in Fig. 6 is characterized by a first peak of about 70mA with a short duration (around 4ns) corresponding to the discharge ignition due to the

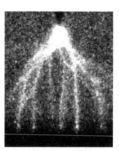

Fig. 7. Time integrated picture of corona discharge in dry air at atmospheric pressure for a time exposure of 10ms: Maximum voltage magnitude=8kV, pulse voltage width=40μs, inter-electrode distance=8mm and pin radius=25μm.

intense ionization processes generated by the high geometric electric field near the pin. This phenomenon can be seen in the first picture of Fig. 8 which shows an intensive spot light around the pin. After this first current peak, as soon as the electron avalanches reach a critical size, the accumulated charge space splits into several streamer heads that begin to propagate towards the cathode (see Fig. 8). During this primary streamer propagation, the corona current begins to steeply increase up to reach a peak value of about 175mA at the streamers arrival at the cathode for a time around 50ns (see Fig. 6). The streamer branches arrive separately at the cathode with an average propagation velocity of about 2.7×10^7 cm.s^{-1}. Above this instant, the corona current peak, after a first decrease due to transition between displacement and conduction currents, slows down during a short duration (around 70ns) corresponding to the secondary streamer propagation. This is then followed by a slower and monotonic fall of the corona current corresponding to the relaxation time that lasts above the 300ns displayed in the time axis of Fig. 6.

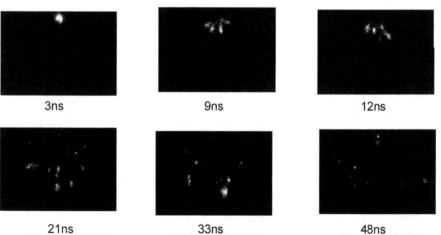

Fig. 8. Time resolved corona discharge pictures at different instants for dry air at atmospheric pressure: Maximum voltage magnitude=8kV, pulse voltage width=40μs, inter-electrode distance=8mm and pin radius=25μm, exposure time = 3ns, reference intensity image=21ns

To summarize, the voltage pulsed corona micro-discharges are characterized by a streamer branching structure and the propagation of multiple primary and secondary streamers. Relations between pulsed and DC voltage conditions can be pointed out using a large width voltage pulse. In this case, several micro-discharges are able to cross the inter-electrode gap during a single voltage pulse. Fig. 9 shows the morphology of the first the 18 corona micro-discharges generated between a pin and a plane using a high voltage pulse of 20ms of duration.

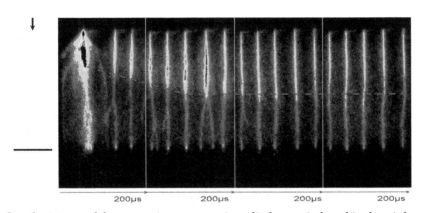

Fig. 9. Streak pictures of the successive corona micro-discharges induced in dry air by a pulse voltage of 20 ms duration and 7.2 kV magnitude (Abahazem et al. 2008).

The branching phenomenon is clearly observed in the first corona micro-discharge with the simultaneous development of a high number of filaments (see the first corona micro-discharge in Fig. 9). Then, during about 400µs, the following discharges present a trunk expansion (shown by the red line in Fig. 9) in front of which a low number of filaments develop. A complete mono-filament structure appears after tens of discharges. Their characteristics are the same as those observed under DC high voltage condition. Thus, the more luminous trails in the discharge pictures of Fig. 9 correspond in fact to the development of the secondary streamers which extend gradually from the point towards the plane. Therefore, the formation of the mono-filament structure can be explained as a result of complex thermal and kinetics memory effects induced in the secondary streamer between each successive discharge.

2.4 Induced neutral gas perturbations

Even if micro-discharges are non thermal plasmas, their propagation can affect the neutral background gas (Eichwald et al. 1997, Ono & Oda b, 2004, Batina et al, 2002). In all cases, the micro-discharges modify the chemical composition of the medium (Kossyi et al. 1992, Eichwald et al. 2002, Dorai & Kushner 2003). In fact, the streamer heads propagate high energetic electrons that create radicals, dissociated, excited and ionized species by collision with the main molecule of the gas. Indeed, we have to keep in mind the low proportion of electrons and more generally of charged particles present in non-thermal plasma. At atmospheric pressure, and in the case of corona micro-discharge, we have about one million of neutral particles surrounding every charged species. Therefore, the collisions charged-neutral particles are predominant. During the discharge phase (which is associated to the current pulse), the radical and excited species are created inside the micro-discharge volume. But during the post-discharge phase (i.e. between two successive current pulses) these active species react with the other molecules and atoms and diffuse in the whole reactor volume. If a gas flow exists, they are also transported by the convective phenomena. However, convective transport can also be induced by the micro-discharges themselves. Indeed, the momentum transfers between heavy charged particles and background gas are able to induce the so called "electric wind". The random elastic collisions between charged and neutral particles directly increase the gas thermal energy. Furthermore, the inelastic processes modify the internal energy of some molecules thus leading to rotational, vibrational and electronic excitations, ionisation and also dissociation of molecular gases. After a certain time, the major part of these internal energy components relaxes into random thermal energy. However, during the lifetime of micro-discharges (some hundred of nanoseconds), only a fraction of this energy, which in fact corresponds mainly to the rotational energy and electronic energy of the radiative excited states, relaxes into thermal form. The other fraction of that energy, which is essentially energy of vibrational excitation, relaxes more slowly (after 10^{-5}s up to 10^{-4}s). The thermal shock during the discharge phase can induce pressure waves and a diminution of the gas density and the vibrational energy relaxation can increase the mean gas temperature (Eichwald et al. 1997). All these complex phenomena induce memory effects between each successive micro-discharge. In fact, the modification of the chemical composition of the gas can favour stepwise ionisation with the pre-excited molecule (like metastable and vibrational excited species), the gas density modification influences all the discharge parameters which are function of the reduced electric field E/N (E being the total electric field and N the background gas density) and the three body reaction that are also function of the gas density. Furthermore, the local temperature increase also modifies the gas reactivity because the efficiency of some reactions depends on the gas temperature following Arrhenius law. Therefore, the complete simulation of micro-discharges has to take into account all these complex phenomena of discharge and gas dynamics.

2.5 The complete micro-discharge model in the hydrodynamics approximation

The complete simulation of the discharge reactor, in complement to experimental studies, can lead to a better understanding of the physico-chemical activity triggered during micro-discharge development and relaxation. Nowadays, in order to take into account the complex energetic, hydrodynamics and chemical phenomena that can influence the corona plasma process, the full simulation of the non thermal plasma reactor can be undertaken by coupling the following models:
- The external electric circuit model,
- the electro-hydrodynamics model,
- the background gas hydrodynamics model including the vibrational excited state evolution,
- the chemical kinetics model,
- and the basic data model which gives the input data for the whole previous models.

Each model gives specific information to the others. For example, the electro-hydrodynamics model gives the morphology of the micro-discharge, the electron density and energy as well as the energy dissipated in the ionized channel by the main charged-neutral elastic and inelastic collision processes. This information is coupled with the external

·lectric circuit model to calculate the micro-discharge impedance needed to follow the inter-·lectrode voltage evolution. On the other hand, the calculated dissipated energy and momentum transfer are included as source terms in the background gas model in order to imulate the induced hydrodynamics phenomena like electric wind, pressure wave ›ropagation, neutral gas temperature increase, etc. The neutral gas hydrodynamics nfluences both the discharge dynamics and the chemical kinetics results. For example, the harged transport coefficients depend on the neutral gas density and some main chemical ·eactions involving neutral species (like the three body reactions) are very dependant on ›oth the gas temperature and density. Finally, the basic data models (Yousfi & ›enabdessadok, 1996, Bekstein et al. 2008, Yousfi et al. 1998, Nelson et al. 2003) give the 1ecessary parameters (such as the convective and diffusive charged and neutral transport ·oefficients, the charged-neutral and neutral-neutral chemical reaction coefficients, the ·raction of the energy transferred to the gas from the elastic and inelastic processes, among ›thers) needed to close the total equation systems.

Γhe electro-hydrodynamics model is an approximation of a more rigorous model. The ‹inetic description based on Boltzmann equations for the charged particles is probably the more rigorous theoretical approach. However the main drawback of the kinetics approach is inked to the treatment of the high number of electrons coming from ionization processes vhich involves huge computation times especially at atmospheric pressure. Therefore, the ·lassical mathematical model used for solving the micro-discharge dynamics is the nacroscopic fluid one also called the hydrodynamics electric model. Up to now, the most ·ommonly used fluid model is the hydrodynamics first order model which involves the first wo moments of Boltzmann equation (i.e the density and the momentum transfer ·onservation equation) for each charged specie coupled with Poisson equation for the space ·harged electric field calculation (Eichwald et al. 1996). In all cases, the momentum equation ·an be simplified into the classical drift-diffusion approximation. The obtained system of ›ydrodynamics equations is then closed by the local electric field approximation which ıssumes that the transport and reaction coefficients of charged particles depend only on the ›ocal reduced electric field E/N. The hydrodynamics approximation is valid as long as the ·elaxation time for achieving a steady state electron energy distribution function is short ·ompared to the characteristic time of the discharge development. At atmospheric pressure ınd because of the high number of collisions, the momentum and energy equilibrium times ıre generally small compared to any macroscopic scale variations of the system. In the ›ydrodynamics approximation, the coupled set of equations that govern the micro-lischarge evolution is the following:

$$\frac{\partial n_c}{\partial t} + \vec{\nabla}.n_c\vec{v}_c = S_c \qquad \forall c \qquad (1)$$

$$n_c\vec{v}_c = \mu_c\vec{E} - \overset{\Rightarrow}{D}_c.\vec{\nabla}n_c \qquad \forall c \qquad (2)$$

$$\varepsilon_0 \Delta V = -\sum_c q_c n_c \qquad (3)$$

$$\vec{E} = -\vec{\nabla}V \qquad (4)$$

These first four equations allow to simulate the behaviour of each charge particle "c" in the micro-discharge (like for example e, N_2^+, O_2^+, O_4^+, O^-, O_2^-, among others). n_c, $\vec{v}_c$, S_c, μ_c, $\overset{\Rightarrow}{D}_c$, q_c are respectively the density, the velocity, the source term, the mobility, the diffusive tensor and the charge of each charge specie "c" involved in the micro-discharge. V and $\vec{E}$ are the potential and the total electric field. The source terms S_c represent for each charge specie the chemical processes (like ionization, recombination, attachment, dissociative attachment, among others) as well as the secondary emission processes (like photo-ionisation and photo-emission from the walls (Kulikovsky, 2000, Hallac et al. 2003, Segur et al. 2006)). The transport equations of charged particles are not only strongly coupled through the plasma reactivity but also through the potential and electric field equations. Indeed, in equation (3) the potential and therefore the electric field in equation (4) are directly dependant on the variation of the density of the charged species, obtained from solution of equations (1)-(2) requiring the knowledge of transport and reaction coefficients that in turn have a direct dependence on the local reduced electric field E/N. Therefore the simulation of micro-discharge dynamics needs fast and accurate numerical solver to calculate the electric field at each time step (especially in regions with high field gradients like near the streamer head and the electrode pin) and also to propagate high density shock wave.

Even if the solution of the first order hydrodynamics model allows a better understanding of the complex phenomena that govern the dynamics of charged particles in micro-discharges, the experimental investigations clearly show that the micro-discharges have an influence on the gas dynamics that can in turn modify the micro-discharge characteristics. It is therefore necessary to couple the electro-hydrodynamics model with the classical Navier-Stockes equations of a compressible and reactive background neutral gas coupled with the conservation equation of excited vibrational energy (Byron et al. 1960, Eichwald et al. 1997).

$$\frac{\partial \rho m_i}{\partial t} + \vec{\nabla}.\rho m_i \vec{v} + \vec{\nabla}.\vec{J}_i = S_i + S_{ic} \qquad \forall i \tag{5}$$

$$\frac{\partial \rho}{\partial t} + \vec{\nabla}.\rho \vec{v} = 0 \tag{6}$$

$$\frac{\partial \rho \vec{v}}{\partial t} + \vec{\nabla}.\rho \vec{v} \vec{v} = -\vec{\nabla}P - \vec{\nabla}.\overset{\Rightarrow}{\tau} + \vec{S}_{qm} \tag{7}$$

$$\frac{\partial \rho h}{\partial t} + \vec{\nabla}.\rho h \vec{v} = \vec{\nabla}.(k\vec{\nabla}T) + \frac{\partial P}{\partial t} + \vec{v}.\vec{\nabla}P + \overset{\Rightarrow}{\tau}:\vec{\nabla}\vec{v} - \vec{\nabla}.\sum_i \vec{J}_i h_i + S_h + \frac{\varepsilon_v}{\tau_v} \tag{8}$$

$$\frac{\partial \varepsilon_v}{\partial t} + \vec{\nabla}.\varepsilon_v \vec{v} = S_v - \frac{\varepsilon_v}{\tau_v} \tag{9}$$

The set of equations (5) to (9) are used to simulate the neutral gas behavior and to follow each neutral chemical species "i" (like N, O, O_3, NO_2, NO, N_2 ($A^3\sum_u^+$), N_2 ($a'^1\sum_u^-$), O_2 ($a^1\Delta g$), among others) that are created during the micro-discharge phase. In equations (5) to (9), ρ is the mass density of the background neutral gas, $\vec{v}$ the gas velocity, P the static pressure and

$\rightarrow$ the stress tensor. For each chemical species "i", m_i is the mass fraction, $\vec{J}_i$ the diffusive ux due to concentration and thermal gradients, S_i the net rate of production per unit olume (due to chemical reactions between neutral species) and S_{ic} simulates the creation f new neutral active species during the discharge phase by electron or ion impacts with the ain molecules of the gas. h is the static enthalpy, T the temperature, k the thermal onductivity and ε_v the vibrational energy. S_h and S_v are the fraction of the total electron ower $\vec{j}.\vec{E}$ transferred during the discharge phase into thermal and vibrational energy. It is enerally assumed that the translational, rotational and electronic excitation energies relax uasi immediately into thermal form and that the vibrational energy stored during the ischarge phases relaxes after a mean delay time τ_v of some tens of micro-seconds. $\vec{S}_{qm}$ is the otal momentum transferred from charged particles to the neutral ones. As already xplained, all the discharge parameters (S_c, μ_c, $\overset{\Rightarrow}{D_c}$, S_{ic}, $\vec{S}_{qm}$, S_h and S_v) are strongly lependent on the reduced electric field (E/N). Therefore the coupling of all the set of quations (1) to (9) for each charged and neutral chemical species will considerably enhance he complexity of the global hydrodynamics model. In fact, each gas density variation can lirectly affect the development of micro-discharges through the reduced electric field ariation.

'inally, the modelling of complex phenomena occurring inside non-thermal reactor filled vith complex gas mixtures needs the knowledge of the electron, the ion and the neutral ransport and reaction coefficients. The charged and neutral particles kinetics model is herefore one of the method in complement to the experimental one that can be used to alculate or complete the set of basic data. Concerning the charged particles, the more ppropriate method to obtain the unknown swarm data is to use a microscopic approach e.g. a Boltzmann's equation solution for the electron data and a Monte Carlo simulation for he ion data) based on collision cross sections (Yousfi & Benabdessadok, 1996, Bekstein et al. '008, Yousfi et al. 1998, Nelson et al. 2003). On the other hand the most commonly used nethod to calculate the neutral swarm data in a gas mixture is the use of the classical kinetic heory of neutral gas mixture (Hirschfielder et al. 1954). The macroscopic charged particles warm data are given over a large range of either the reduced electric field or the mean .lectron energy. The whole set of data includes:

The macroscopic transport coefficients like mobility, longitudinal and transversal diffusion coefficients,

the reaction coefficients like ionization, attachment, dissociation, radiative or metastable electronic excitation coefficients,

the mean electron energy exchange frequencies of the elastic, inelastic and super-elastic processes,

and the mean electron momentum exchange frequency (if the classical drift diffusion approximation is not assumed valid)

[he calculation of the scalar (e.g. ionization or attachment frequencies), vectorial (drift relocity), and tensorial (diffusion coefficients) hydrodynamics electron and ion swarm 'arameters in a gas mixture, needs the knowledge of the elastic and inelastic electron-nolecule and ion-molecule set of cross sections for each pure gas composing the mixture. 'ach collision cross section set involves the most important collision processes that either affect the charged species transport coefficients or are needed to follow the charged specie chemical kinetics and energy or momentum exchange. For example, in order to calculate the macroscopic electron swarm parameters in water vapor, 21 collision cross sections must be known involving the rotational, the vibrational and the electronic excitation processes a well as the ionization, the dissociative attachment and the superelastic processes.

One of the main difficulties is to validate for each pure gas that compose the mixture the chosen set of cross sections. To do that, a first reliable set of electron-molecule and ion molecule cross section for each individual neutral molecule in the gas mixture must be known Then, in order to obtain the complete and coherent set of cross sections, it is necessary to adjust this first set of cross sections so as to fit experimental macroscopic coefficients with the calculated ones estimated from either a Boltzmann's equation solution or a Monte Carlo simulation. The obtained solution is certainly not unique but as the comparisons concern several kinds of swarm macroscopic parameters having different dependencies on cross sections (ionization or attachment coefficient, drift velocity, transverse or longitudinal diffusion coefficient) over a wide range of reduced electric field or mean electron energy, most of the incoherent solutions are rejected. Finally, when the sets of cross section are selected for each pure gas, they can be used to calculate with a Bolzmann's equation solution or a Monte Carlo simulation the macroscopic charged species transport and reaction parameters whatever the proportion of the pure gas in the background gas mixture.

2.6 Summary

Micro-discharges are characterized by the development of primary and secondary streamers. As a function of the high voltage applied on the small curvature electrode (DC or pulse), the micro-discharges show either a mono-filament or a large branching structure The passage from multi-filaments to mono-filament structure can be observed if a sufficiently large high voltage pulse is applied. The transition can be explained through the memory effects accumulated during the previous discharge. The primary streamers propagate fast ionization waves characterized by streamer heads in which the electric field is high enough to generate high energetic electrons like in an electron gun. The streamer head propagates a high charge quantity toward the inter-electrode gap. The micro-plasmas are generated behind the streamer heads. They are small conductive channels that connect the streamer head to the electrode stressed by the high voltage. The primary streamers are then followed by a secondary streamer which is characterized by an electric field extension that ensures the transition between the displacement current and the conductive one when the primary streamer arrives on the cathode. Both primary and secondary streamers create radicals and excited species by electron-molecule impacts. The elastic and inelastic energy transfers generate a chemical activity, a thermal energy increase of the gas and a neutral gas dynamics. To better understand all these complex phenomena, a hydrodynamics model can be used based on conservation equations of charged and neutral particles coupled to Poisson equation for the electric field calculation.

3. Chemical and hydrodynamics activation of gases using corona micro-discharge

3.1 Introduction

During the past two decades several studies have shown that non-thermal plasmas reactor working in ambient air are very efficient sources of active species like charged particles, adicals and excited species. In fact, and as already explained in the previous sections, in the non-thermal plasma reactor, the majority of the injected electrical energy goes into the generation of energetic electrons, rather than into gas heating. The energy in the micro-plasma is thus directed preferentially to electron-impact dissociation, excitation and ionization of the background gas to generate active species that, in turn, induce the chemical activation of the medium. As a consequence, the non-thermal plasma reactors at atmospheric pressure are used in many applications such as flue gas pollution control (Fridman et al., 2005, Urashima et Chang, 2010), ozone production (Ono & Oda b, 2004), surface decontamination (Clement et al., 2001, Foest et al., 2005) and biomedical field (Laroussi, 2002, Villeger et al., 2008, Sarrette et al., 2010). For many applications, particularly in the removal of air pollutants, decontamination or medicine field, the non-thermal plasma approach would be most appropriate because of its energy selectivity and its capability for simultaneous treatment of pollutants, bacteria or cells for example.

In micro-discharges the active species are created by energetic electrons during the primary and the secondary streamer propagation that last some hundred of nanoseconds. Despite these very fast phenomena, the energy transferred to the gas can initiate shock waves starting from the stressed high voltage electrodes. Furthermore, a part of the electronic energy is stored in the vibrational energy that relaxes in thermal form after some tens of microseconds. Anyway, it is worth to notice, that all the initial energy (chemical, thermal, among others) is transferred inside a very thin discharge filament i.e. in a very small volume compared with the volume of the plasma reactor. Therefore, the efficiency of the processes is correlated to the radical production efficiency during the discharge phase, the number of micro-discharges that cross the inter-electrode gap, the repetition frequency of the discharge and how the radicals are diffused and transported from the micro-discharge towards the whole reactor volume.

In the following sections, the discharge and the post-discharge phase are simulated using the hydrodynamics models presented in section 2.5 in the case of a DC positive pin-to-plan corona reactor in dry air at atmospheric pressure.

3.2 Discharge phase simulation

The simulation conditions are described in detail in reference (Eichwald et al. 2008) as well as the used numerical methods and boundary conditions. To summarize, a DC high voltage of 7.2kV is applied on the pin of a pin-to-plane reactor filled with dry air at atmospheric pressure. The inter-electrode gap is of 7mm, the pin radius is equal to 25µm and photo-ionisation phenomenon is taken into account in the simulation. Results in Fig. 10 and 11 are obtained by coupling equations (1) to (4) for electrons, two negative ions (O$^-$ and O$_2^-$), four positive ions (N$_2^+$, O$_2^+$, N$^+$ and O$^+$) and two radical atoms (O, N) reacting following 10 selected reactions. Because of the time scale of the discharge phase (some hundred of nanoseconds), the radical atoms and the main neutral molecules (N$_2$ and O$_2$) are supposed to remain static during the discharge phase simulation. Fig. 10 shows the reduced electric field (E/N) expressed in Td (1Td=10^{-21} Vm^2 so that 500Td at atmospheric pressure is equivalent to an electric field of 12MVm^{-1}). When the high voltage is applied to the pin, some seed electrons are accelerated in the high geometric electric field around the pin. A luminous spot is observed experimentally near the pin thus indicating the formation of excited species due to a high electronic energy. On can notice that the electrons move towards the pin. Furthermore, the electrons gain sufficient energy to perform electronic avalanches and a plasma spot is created just around the pin. The plasma is a quasi neutral electric gas in which the quantity of negative and positive species is quasi similar. Nevertheless, as the mobility of electrons is much higher than those of positive ions, the electric neutrality of the initial plasma spot is perturbed just in front of the pin. Indeed, the electrons are absorbed by the positive anodic pin while the positive ions remain quasi static due to their mass inertia. A positive charged space is formed and the electric field is no more at his maximum on the pin but just in front of it. This situation can be seen at time t=20ns in the first picture of Fig. 10.

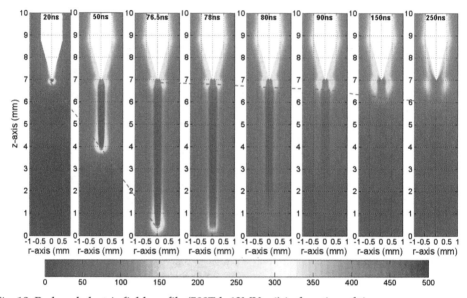

Fig. 10. Reduced electric field profile (500Td=12MVm⁻¹) in function of time

A streamer head is created that propagates from the pin towards the plane. This streamer head can be interpreted as the propagation of a positive charge space shock wave. At each time of its propagation, new seed electrons are created in front of the streamer head by photo-ionisation processes. These electrons are accelerated in the high electric field and their energy is high enough to ionize, dissociate and excite the main molecules of the gas. When the electrons have crossed the streamer head they drift towards the pin inside a small conductive plasma channel that connects the streamer head to the pin. A micro-plasma is formed behind the streamer head and is constricted by a cylinder of space charged electric field. A quasi-homogeneous small value of electric field is maintained inside the micro-plasma in order to allow the drift of electrons from the streamer head to the pin that ensures the continuity of the total current density. The time laps needed for the streamer head to cross the inter-electrode gap is associated with the primary streamer propagation of the micro-discharge phase. The streamer head propagates a charge quantity which is absorbed by the cathode plane as soon as it arrives on the cathode plane. It results to the first current peak observed in Fig. 4 for a high voltage DC condition. The first red dashed curve in Fig. 10 follows the trail left by the high electric field of the streamer head. Its shape corresponds to the luminous trail observed by streak camera shown in Fig. 3. When the primary streamer rrives at the cathode plane, a secondary streamer starts its propagation from the pin. The secondary streamer is an electric field plateau extension of value of about 100Td. This extension ensures the continuity of the total current when the total charge space transported by the streamer head is absorbed at the cathode (Eichwald et al. 2008, Bastien & Marode, 985). The second red dashed curve follows the plateau extension. The evolution of the luminous trail left by the secondary streamer shown in Fig. 3 is due to the excited species created by the energetic electrons inside the secondary streamer expansion. Fig. 11 shows the radical O density after 150ns. The simulation indicates that about 70% of the radical O is produced inside the secondary streamer by dissociative collisions between electrons and O_2 molecules in reaction $e + O_2 \rightarrow O + O$. The concentration of O radical is also high near the cathode plane due to a higher electric field magnitude inside the streamer head when it reaches the cathode.

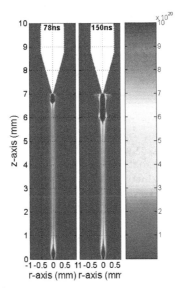

Fig. 11. O radical profile (m^{-3})

The effect on the neutral gas dynamics induced by the micro-discharge propagation is shown in Fig. 12 and 13. Fig. 12 shows the temperature profile of the background neutral gas at 0.1µs (=100ns) and 0.3µs (=300ns). Fig. 13 shows the pressure profile from 0.1µs to 4µs. The gas temperature just on the pin reaches some thousands of Kelvin but a mean value of about 700°K is obtained around the point. This value is coherent with experimental results obtained under very similar condition (Spyrou et al., 1992). The thermal shock creates high pressure gradients (see Fig. 13 at 0.1µs) and induces the gas expansion (see Fig. 13). Due to the inertia principle, the mass density near the point decreases more gradually in a time scale greater than the temperature increase. The gas expansion is characterised by a cylindrical and a spherical shock wave (see Fig. 13 from 0.3 to 0.9µs). Indeed, the initial pressure gradients (which induced the gas motion) follow the temperature ones which are constricted along the axis and inside the micro-plasma channel. We therefore observe a cylindrical pressure wave (represented by two vertical lines in the flat pressure mappings of Fig. 13) that propagates from the streamer axis towards the whole domain. The complex

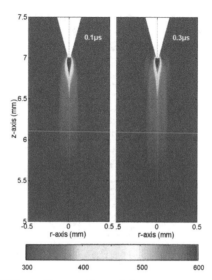

Fig. 12. Gas temperature (°K) profile near the point

structure of the pressure gradients near the point induces a spherical pressure wave superimposed to the cylindrical one. Such kind of spherical pressure waves were already observed experimentally (Ono & Oda b, 2004) using the laser Schlieren method. Furthermore, the simulation shows that the spherical shock wave propagates at the speed of sound as in the case of experimental work (Ono & Oda b, 2004).

In this kind of simulation the effects of temperature and gas density variations on the streamer development are not taken into account. However, it should be in further works because if the gas density varies it will modify the reduced electric flied (E/N) and therefore the behaviour of the charged particles whose properties (like mobility, ionisation frequency,…) completely depend on the reduced electric field. Nevertheless, the previous results are able to give the initial profiles of all the source terms needed to simulate the post-discharge phase evolution.

3.3 Post-discharge phase simulation

The discharge phase simulation gives very clear information on the gas dynamics and the spatio-temporal evolution of each active species of the background gas mixture. However, the time and space scales between the discharge phase and the post-discharge phase are completely different. Indeed, the micro-discharge generated micro-plasma in some hundred of nanoseconds while post-discharge phase must be considered with centimetre scale and milliseconds time laps. A complete simulation of both coupled phenomena for multi-pin reactor needs therefore adaptive meshes from micrometer to centimetre scale and also adaptive time scale from picoseconds (in order to follow the nano-scale discharge phenomena) up to fraction of milliseconds. This means a large number of discrete spatial cells and a huge computing time. In order to overcome these difficulties, on can assume that the effects of the discharges on the background gas can be simulated by locally injected inside the micro-discharge volumes and only during the discharge phase, average source terms estimated from the complete discharge phase model.

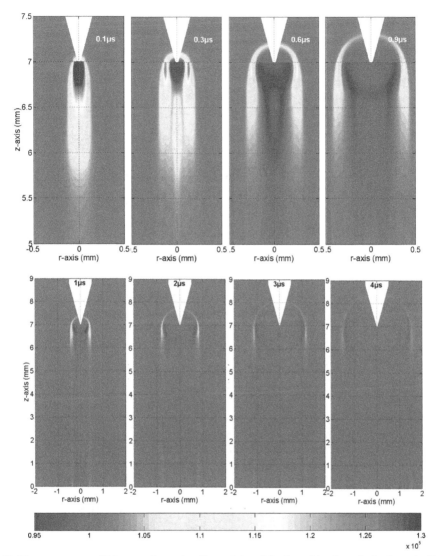

Fig. 13. Pressure wave (Pa) near the point (from 0.1 to 0.9µs) and in the whole domain (from 1 to 4µs)

As an example, let us suppose the multi-pin reactor described in Fig. 14. The domain is divided with square structured meshes of 50µm×50µm size. A DC high voltage of 7.2kV is applied on the pins. During each discharge phase, monofilament micro-discharges are created between each pin and the plane with a natural frequency of 10kHz. The micro-discharges have an effective diameter of 50µm which correspond to the size of the chosen cells. Therefore, it is possible to inject in the cells located between each pin and the plane specific profiles of active source species and energy that will correspond the micro-discharge effects.

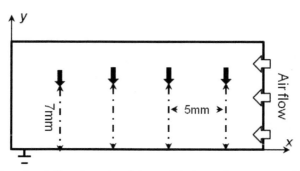

Fig. 14. 2D Cartesian simulation domain of the multi-pin to plane corona discharge reactor.

As an example, consider equation (5) of section 2.5 applied to O radical atoms ('i"=O).

$$\frac{\partial \rho m_O}{\partial t} + \vec{\nabla}.\rho m_O \vec{v} + \vec{\nabla}.\vec{J}_O = S_O + S_{Oc}$$

The challenge is to correctly estimate the source term S_{Oc} inside the volume of each micro-discharge. As the radial extension of the micro-discharges is equal to the cell size, the source term between each pin and the plane depends only on variable z. The average source term responsible of the creation of O radical during the discharge phase is therefore expressed as follow:

$$S_{Oc}(z) = \frac{1}{r_d}\frac{1}{t_d}\int_0^{r_d}\int_0^{t_d} s_{Oc}(t,r,z)\,dt\,dr \tag{10}$$

t_d is the effective micro-discharge duration, r_d the effective micro-discharge radius and $s_{Oc}(t,r,z)$ the source terms (m^{-3}s^{-1}) of radical production during the discharge phase (i.e. $k(E/N)n_e n_{O2}$ for reaction $e + O_2 \rightarrow O + O$ where $k(E/N)$ is the corresponding reaction coefficient). All the data in equation (10) come from the complete simulation of the discharge phase. In the present simulation conditions, specific source terms are calculated for 5 actives species that are created during the discharge phase (N$_2$(A$^3\Sigma_u^+$), N$_2$(a'$^1\Sigma_u^-$), O$_2$(a$^1\Delta$g), N and O).

The energy source terms in equations (8) and (9) are estimated using equations (11) and (12):

$$S_h(z) = \rho C_p \frac{1}{r_d}\frac{1}{t_p^2}\int_0^{r_d}\int_0^{t_p} T(t_p,r,z)\,dt\,dr \tag{11}$$

$$S_v(z) = \frac{1}{r_d}\frac{1}{t_d}\int_0^{r_d}\int_0^{t_d} f_v \vec{J}.\vec{E}\,dt\,dr \tag{12}$$

In equation (12), $\vec{J}.\vec{E}$ is the total electron density power gained during the discharge phase and f_v the fraction of this power transferred into vibrational excitation state of background gas molecules. One can notice the specificity of equation (11) related with the estimation of the direct random energy activation of the gas. In this equation, t_p is the time scale of the pressure wave generation rather than the micro-discharge duration t_d. In fact, during the post-discharge phase, the size of discrete cells is not sufficiently small to follow the gradients of pressure wave generated by thermal shock near the point (see Fig. 13). However, pressure waves transport a part of the stored thermal energy accumulated around each pin. From 0.1μs to 0.3μs, the gas temperature on the pins decreases from about 3000°K down to about 1200°K. After this time, the temperature variation in the micro-discharge volume is less affected by the gas dynamics. The diffusive phenomena become predominant. Therefore, taken into account the mean energy source term at time t_d will overestimate the temperature enhancement on the pins during the post-discharge phase simulation. As a consequence, the time t_p is chosen equal to 300ns i.e. after the pressure waves have left the micro-discharge volume.

As an example, Fig. 15 shows the temperature profile obtained at $t=t_p$ just after the first discharge phase. The results were obtained using the Fluent Sofware in the simulation conditions described in Fig. 14. As expected and just after the first discharge phase, the enhancement of the gas temperature is confined only inside the micro-plasma filaments located between each pin and the plane. The temperature profile along the inter-electrode gap is very similar to the one obtained by the complete discharge phase simulation (see Fig. 12). It is also the case for the active source terms species. Fig. 16 shows at time $t=t_d$, the axial profile of some active species that are created during the discharge phase. The curves of the discharge model represent the axial profile density averaged along the radial direction. In

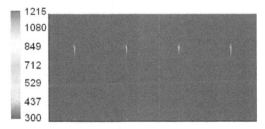

Fig. 15. Gas temperature profile after the first discharge phase at $t=t_p = 300$ns.

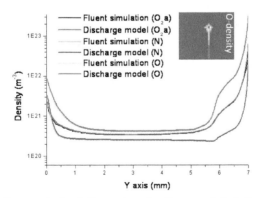

Fig. 16. Comparison of numerical solutions given by the completed discharge and Fluent models at $t_d=150$ ns for O, N and O_2 ($a^1\Delta g$) densities. The zoom box shows, as an example, the O radical profile near a pin.

the case of the O radical, the density profile of Fig. 11 was averaged along the radial direction until $r_d=50\mu m$ and drawn in Fig. 16 with the magenta color. The light blue color curve represents the O radical profile obtained with the Fluent Software when the specific source term profile $S_{Oc}(z)$ is injected between a pin and the cathode plane in the simulation conditions of Fig. 14.

In the following results, the complete simulation of the successive discharge and post-discharge phases involves 10 neutral chemical species (N, O, O_3, NO_2, NO, O_2, N_2, N_2 ($A^3\Sigma_u^+$), N_2 ($a'^1\Sigma_u^-$) and O_2 ($a^1\Delta g$)) reacting following 24 selected chemical reactions. The pin electrodes are stressed by a DC high voltage of 7.2kV. Under these experimental conditions the current pulses appear each 0.1ms (i.e. with a repetition frequency of 10KHz). It means that the previous described source terms are injected every 0.1ms during laps time t_d or t_p and only locally inside the micro-plasma filament located between each pin and the plane. The lateral air flow is fixed with a neutral gas velocity of 5m.s^{-1}.

Pictures in Fig. 17 show the cartography of the temperature and of the ozone density after 1ms (i.e. after 10 discharge and post-discharge phases). One, two, three or four pins are stressed by the DC high voltage. Pictures (a) show that for the mono pin case, the lateral air flow and the memory effect of the previous ten discharges lead to a wreath shape of the space distribution of both the temperature and the ozone density.

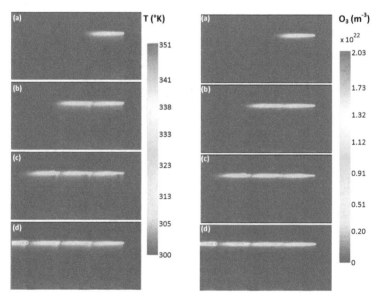

Fig. 17. Temperature and ozone density profile at 1ms i.e. after ten discharge and post-discharge phases. The number of high voltage pin is respectively (a) one, (b) two, (c) three and (d) four. The lateral air flow is 5m.s^{-1}.

The temperature and the ozone maps are very similar. Indeed, both radical and energy source terms are higher near the pin (i.e. inside the secondary streamer area expansion as it was shown in section 3.2). Furthermore, the production of ozone is obviously sensitive to the gas temperature diminution since it is mainly created by the three body reaction $O+O_2+M \rightarrow O_3+M$ (having a reaction rate inversely proportional to gas temperature).

For more than one pin, the temperature and ozone wreaths interact each other and their superposition induce locally a rise of both the gas temperature and ozone density (see Fig. 17). The local maximum of temperature is around 325K for one pin case and increases up to 350K for four anodic pins.

The average temperature in the whole computational domain remains quasi constant and the small variations show a linear behavior with the number of anodic pins. The same linear tendency is observed for the ozone production in Fig. 18. After 1ms, and for the four pins case, the mean total density inside the computational domain reaches 4×10^{14} cm^{-3}.

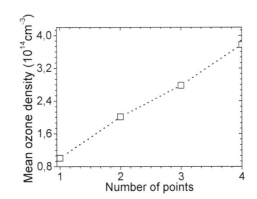

Fig. 18. Mean ozone density increase inside the computational domain of Fig. 14 as a function of the number of pins

3.4 Summary

The complete simulation of all the complex phenomena that are triggered by micro-discharges in atmospheric non thermal plasma was found to be possible not as usually done in the literature only for 0D geometry but also in multidimensional geometry. In DC voltage conditions, a specific first order electro-hydrodynamics model was used to follow the development of the primary and secondary streamers in mono pin-to-plane reactor. The simulation results reproduce qualitatively the experimental observations and are able to give a full description of micro-discharge phases. Further works, already undertaken in small dimensions or during the first instants of the micro-discharge development (Pancheshnyi 2005, Papageorgiou et al. 2011), have to be achieved in 3D simulation in order to describe the complex branching structure for pulsed voltage conditions. Nevertheless, the micro-discharge phase simulation gives specific information about the active species profiles and density magnitude as well as about the energy transferred to the background gas. All these parameters were introduced as initial source terms in a more complete hydrodynamics model of the post-discharge phase. The fist obtained results show the ability of the Fluent software to solve the physico-chemical activity triggered by the micro-discharges.

4. Conclusion

The present chapter was devoted to the description of the hydrodynamics generated by corona micro-discharges at atmospheric pressure. Both experimental and simulation tools have to be exploited in order to better characterise the strongly coupled behaviour of micro-discharges dynamics and background gas dynamics. The experimental devices have to be very sensitive and precise in order to capture the main characteristics of nanosecond phenomena located in very thin filaments of micro scale extension. However, the recent evolution of experimental devices (ICCD or streak camera, DC and pulsed high voltage supply, among others) allow to better understand the physics of the micro-discharge. Furthermore, recent simulation of the micro-discharges involving the discharge and post-discharge phase in multidimensional dimension was found to give precise information about the chemical and hydrodynamics activation of the background gas in an atmospheric non-thermal plasma reactor. These kinds of simulation results, coupled with experimental investigation, can be used in future works for the development of new design of plasma reactor very well adapted to the studied application either in the environmental field or biomedical one.

5. Acknowledgment

All the simulations were performed using the HPC resources from CALMIP (Grant 2011-[P1053] - www.calmip.cict.fr/spip/spip.php?rubrique90)

6. References

Abahazem, A.; Merbahi, N.; Ducasse, O.; Eichwald, O. & Yousfi, M. (2008), Primary and secondary streamer dynamics in pulsed positive corona discharges, *IEEE Transactions on Plasma Science* , Vol. 36, No. 4, pp. 924-925

Bastien, F. & Marode, E. (1985), Breackdown simulation of electronegative gases in non-uniform field, *Journal of Physics D: Applied Physics*, Vol. 18, pp. 377-393

Batina, J.; Noel, F.; Lachaud, S. ; Peyrous, R. & Loiseau, J. F. (2001) Hydrodynamical simulation of the electric wind in a cylindrical vessel with positive point-to-plane device, *Journal of Physics D: Applied Physics*, Vol. 34, pp. 1510–1524

Bekstein, A.; Benhenni, M.; Yousfi, M.; Ducasse, O. & Eichwald, O. (2008), Ion swarm data of N_4^+ in N_2, O_2 , and dry air for streamer dynamics simulation , *European Physics Journal Applied Physics*, Vol. 42, pp. 33-40

Briels, T. M. P.; Kos J.; van Veldhuizen E. M. & Ebert, U. (2006), Circuit dependence of the diameter of pulsed positive streamers in air, Journal of Physics D: Applied Physics, Vol. 39, pp. 5201–5210

Byron, R.; Stewart, W. E.; Lightfoot E. N. (1960) Transport Phenomena, John Wiley & Sons

Clement, F.; Held, B.; Soulem N. & Spyrou N. (2001). Polystyrene thin films treatment under DC pulsed discharges conditions in nitrogen, *The European Physical Journal, Applied Physics*, Vol. 13, pp. 67-75

Dubois, D.; Merbahi, N.; Eichwald, O.; Yousfi, M.; Ducasse, O. & Benhenni, M. (2007), Electrical analysis of DC positive corona discharge in air and N_2, O_2 and CO_2 mixtures, *Journal of Applied Physics*, Vol. 101, Issue 5, pp. 053304-053304-9

Dorai R. & Kushner M. (2003) Consequences of unburned hydrocarbons on microstreamer dynamics and chemistry during plasma remediation of NOx using dielectric barrier discharges *Journal of Physics D: Applied Physics*,. Vol. 36, pp. 1075–1083

Eichwald, O.; Ducasse, O.; Merbahi, N.; Yousfi, M. & Dubois, D. (2006), Effect of order fluid models on flue gas streamer dynamics", *Journal of Physics D: Applied Physics*, Vol. 39, pp. 99–107

ichwald, O.; Yousfi M.; Hennad A.& Benabdessadok M. D. (1997) Coupling of chemical kinetics, gas dynamics, and charged particle kinetics models for the analysis of NO reduction from flue gases, *Journal of Applied Physics*, Vol. 82, No. 10, pp. 4781-4794

ichwald, O.; Ducasse, O.; Dubois, D.; Abahazem, A.; Merbahi, N.; Benhenni M. & Yousfi, M. (2008) Experimental analysis and modelling of positive streamer in air: towards an estimation of O and N radical production, J. Phys. D: Appl. Phys., Vol. 41 234002 (11pp)

ichwald, O., Guntoro, N. A.; Yousfi, M. & Benhenni M. (2002) Chemical kinetics with electrical and gas dynamics modelization for NOx removal in an air corona discharge *Journal of Physics D: Applied Physics*, Vol. 35, pp. 439-450

oest, R.; Kindel, E.; Ohl, A.; Stieber, M. & Weltmann, K. D. (2005) Non-thermal atmospheric pressure discharges for surface modification Plasma Physics Controlled Fusion, Vol. 47, B525–B536

ridman A.; Chirokov, A. & Gutsol, A. (2005), TOPICAL REVIEW, Non-thermal atmospheric pressure discharges, *Journal of Physics D: Applied Physics*, Vol. 38, R1-R24

allac, A.; Georghiou, G. E. & Metaxas, A; C. (2003) Secondary emission effects on streamer branching in transient non-uniform short-gap discharges, *Journal Of Physics D: Applied Physics*, Vol. 36 pp. 2498–2509

irschfielder, J. O.; Curtiss, F. E. and Bird R. B. (1954) Molecular theory of gases and liquids, Wiley, New York

ossyi, I. A. ; Yu Kostinsky, A. ; Matveyev A. A. & Silakov V. P., (1992), Kinetic scheme of the non equilibrium discharge in nitrogen-oxygen mixtures, *Plasma Sources Sciences and Technologies*, Vol.1, pp. 207-220

ulikovsky, A. A. (2000), The role of photoionisation in positive streamer dynamics, *Journal of Physics D: Applied Physics*, Vol. 33, pp. 1514-1524

aroussi, M. (2002). Nonthermal decontamination of biological media by atmospheric-pressure plasmas: review, analysis and prospects, *IEEE Transactions on Plas. Science*, Vol. 30, No.4, pp. 1409-1415

oeb, L. B. (1961) Basic processes of gaseous electronics, University of California Press, Berkeley.

oeb, L. B. (1965) Electrical coronas, University of California Press, Berkeley.

Marode, E. (1975), The mechanism of spark breakdown in air at atmospheric pressure between a positive point an a plane. I. Experimental: Nature of the streamer track, *Journal of Applied Physics*, Vol. 46, No. 5, pp. 2005-2015

Nelson, D.; Benhenni, M.; Eichwald, O. & Yousfi, M. (2003), Ion swarm data for electrical discharge modeling in air and flue gas, *Journal of Applied Physics*, Vol. 94, pp. 96-103

no, R. & Oda, T. a (2004). Spatial distribution of ozone density in pulsed corona discharges observed by two-dimensional laser absorption method, *Journal of Physics D: Applied Physics*, Vol. 37, pp. 730-735.

no, R. & Oda, T. b (2004) Visualization of Streamer Channels and Shock Waves Generated by Positive Pulsed Corona Discharge Using Laser Schlieren Method, Japanese Journal of Applied Physics, Vol. 43, No. 1, 2004, pp. 321–327

ancheshnyi, S. (2005), Role of electronegative gas admixtures in streamer start, propagation and branching phenomena, *Plasma Sources Science and Technology*, Vol. 14, pp. 645–653

Papageorgiou, L., Metaxas, A. C. & Georghiou, G. E. (2011), Three-dimensional numerical modelling of gas discharges at atmospheric pressure incorporating photoionization phenomena, *Journal of Physics D: Applied Physics*, Vol. 44, 045203 (10pp).

Penetrante, B. M. & Schultheis, S. E. (1993), Nonthermal Plasma Techniques for Pollution Control, Part A&B, Editors, NATO ASI Series Vol. G 34, Springer-Verlag Karlsruhe.

Sarrette, J. P. ; Cousty, S. ; Merbahi, N. ; Nègre-Salvayre, A. & F. Clément (2010) Observation of antibacterial effects obtained at atmospheric and reduced pressures in afterglow conditions, *European Physical Journal. Applied. Physics*, Vol. 49 13108

Segur, P.; Bourdon, A.; Marode, E., Bessieres D. & Paillol J. H. (2006) The use of an improved Eddington; approximation to facilitate the calculation of photoionization in streamer discharges, Plasma Sources Sciences and Technologies, Vol. 15, pp. 648–660

Sigmond, R. S. (1984), The residual streamer channel: Return strockes and secondary streamers, *Journal of Applied Physics*, Vol. 56, No. 5, pp. 1355-1370

Spyrou, N.; Held, B.; Peyrous, R.; Manassis, Ch. & Pignolet, P. (1992) Gas temperature in a secondary streamer discharge: an approach to the electric wind, Journal of Physics D: Applied Physics, Vol. 25, pp. 211-216

van Veldhuizen, E. M. & Rutgers, W. R. (2002), Pulsed positive corona streamer propagation and branching, *Journal of Physics D: Applied Physics*, Vol. 35, pp. 2169–2179

Urashima K. & J. S. Chang (2010) Removal of Volatile Organic Compounds from Air Streams and Industrial Flue Gases by Non-Thermal Plasma Technolog, *IEEE Transactions on Dielectrics and Electrical Insulation*, Vol. 7 No. 5, pp. 602-614

Villeger, S.; Sarrette, J. P.; Rouffet, B. ; Cousty S. & Ricard A. (2008), Treatment of flat and hollow substrates by a pure nitrogen flowing post discharge: Application to bacterial decontamination in low diameter tubes, *European Physical Journal. Applied. Physics*, Vol. 42, pp. 25-32.

Winands, G.; Liu, Z.; Pemen, A.; van Heesch, E.; Yan, K. & van Veldhuizen, E. (2006), Temporal development and chemical efficiency of positive streamers in a large scale wire-plate reactor as a function of voltage waveform parameters, *Journal of Physics D: Applied Physics*, Vol. 39, pp. 3010–3017

Yousfi, M. & Benabdessadok, M. D. (1996), Boltzmann equation analysis of electron-molecule collision cross sections in water vapor and ammonia, *Journal of Applied Physics*, Vol. 80, pp. 6619-6631

Yousfi, M.; Hennad, A. & Eichwald, O. (1998), Improved Monte Carlo method for ion transport in ion-molecule asymmetric systems at high electric fields, *Journal of Applied Physics*, Vol. 84, No. 1, pp. 107-104

3

Hydrodynamics on Charged Superparamagnetic Microparticles in Water Suspension: Effects of Low-Confinement Conditions and Electrostatics Interactions

P. Domínguez-García[1] and M.A. Rubio[2]

[1]Dep. Física de Materiales, UNED, Senda del Rey 9, 28040. Madrid
[2]Dep. Física Fundamental, UNED, Senda del Rey 9, 28040. Madrid
Spain

1. Introduction

The study of colloidal dispersions of micro-nano sized particles in a liquid is of great interest for industrial processes and technological applications. The understanding of the microstructure and fundamental properties of this kind of systems at microscopic level is also useful for biological and biomedical applications.

However, a colloidal suspension must be placed somewhere and the dynamics of the micro-particles can be modified as a consequence of the confinement, even if we have a low-confinement system. The hydrodynamics interactions between particles and with the enclosure's wall which contains the suspension are of extraordinary importance to understanding the aggregation, disaggregation, sedimentation or any interaction experienced by the microparticles. Aspects such as corrections of the diffusion coefficients because of a hydrodynamic coupling to the wall must be considered. Moreover, if the particles are electrically charged, new phenomena can appear related to electro-hydrodynamic coupling.

Electro-hydrodynamic effects (Behrens & Grier (2001a;b); Squires & Brenner (2000)) may have a role in the dynamics of confined charged submicron-sized particles. For example, an anomalous attractive interaction has been observed in suspensions of confined charged particles (Grier & Han (2004); Han & Grier (2003); Larsen & Grier (1997)). The possible explanation of this observation could be related with the distribution of surface's charges of the colloidal particles and the wall (Lian & Ma (2008); Odriozola et al. (2006)). This effect could be also related to an electrostatic repulsion with the charged quartz bottom wall or to a spontaneous macroscopic electric field observed on charged colloids (Rasa & Philipse (2004)). In this work, we are going to describe experiments performed by using magneto-rheological fluids (MRF), which consist (Rabinow (1948)) on suspensions formed by water or some organic solvent and micro or nano-particles that have a magnetic behaviour when a external magnetic field is applied upon them. Then, these particles interact between themselves forming aggregates with a shape of linear chains (Kerr (1990)) aligned in the direction of the magnetic field. When the concentration of particles inside the fluid is high enough, this microscopic behaviour turns to significant macroscopic consequences, as an one million-fold increase in the viscosity of the fluid, leading to practical and industrial applications, such as mechanical devices of different types (*Lord Corporation, http://www.lord.com/* (n.d.); Nakano & Koyama (1998); Tao (2000)). This magnetic particle technology has been revealed as useful in other fields such as microfluidics (Egatz-Gómez et al. (2006)) or biomedical techniques (Komeili (2007); Smirnov et al. (2004); Vuppu et al. (2004); Wilhelm, Browaeys, Ponton & Bacri (2003); Wilhelm et al. (2005)).

In our case, we investigate the dynamics of the aggregation of magnetic particles under a constant and uniaxial magnetic field. This is useful not only for the knowledge of aggregation properties in colloidal systems, but also for testing different models in Statistical Mechanics. Using video-microscopy (Crocker & Grier (1996)), we have measured the different exponents which characterize this process during aggregation (Domínguez-García et al. (2007)) and also in disaggregation (Domínguez-García et al. (2011)), i.e., when the chains vanishes as the external field is switched off. These exponents are based on the temporal variation of the aggregates' representative quantities, such as the size s or length l. For instance, the main dynamical exponent z is obtained through the temporal evolution of the chains length $s \sim t^z$. Our experiments analyse the microstructure of the suspensions, the aggregation of the particles under external magnetic fields as well as disaggregation when the field is switched off. The observations provide results that diverge from what a simple theoretical model says. These differences may be related with some kind of electro-hydrodynamical interaction, which has not been taken into account in the theoretical models.

In this chapter, we would like first to summarize the basic theory related with our system of magnetic particles, including magnetic interactions and Brownian movement. Then, hydrodynamic corrections and the Boltzmann sedimentation profile theory in a confined suspension of microparticles will be explained and some fundamentals of electrostatics in colloids are explained. In the next section, we will summarize some of the most recent remarkable studies related with the electrostatic and hydrodynamic effects in colloidal suspensions. Finally, we would like to link our findings and investigations on MRF with the theory and studies explained herein to show how the modelization and theoretical comprehension of these kind of systems is not perfectly understood at the present time.

2. Theory

In this section, we are going to briefly describe the theory related with the main interactions and effects which can be suffered by colloidal magnetic particles: magnetic interactions, Brownian movement, hydrodynamic interactions and finally electrostatic interactions.

2.1 Magnetic particles

By the name of "colloid" we understand a suspension formed by two phases: one is a fluid and another composed of mesoscopic particles. The mesoscopic scale is situated between the tens of nanometers and the tens of micrometers. This is a very interesting scale from a physical point of view, because it is a transition zone between the atomic and molecular scale and the purely macroscopic one.

When the particles have some kind of magnetic property, we are talking about magnetic colloids. From this point of view, two types of magnetic colloids are usually considered: ferromagnetic and magneto-rheologic. The ferromagnetic fluids or ferrofluids (FF) are colloidal suspensions composed by nanometric mono-domain particles in an aqueous or organic solvent, while magneto-rheological fluids (MRF) are suspensions of paramagnetic micro or nanoparticles. The main difference between them is the permanent magnetic moment

Hydrodynamics on Charged Superparamagnetic Microparticles in Water Suspension: Effects of Low-Confinement Conditions and Electrostatics Interactions

55

of the first type: while in a FF, magnetic aggregation is possible without an external magnetic field, this does not occur in a MRF. The magnetic particles of a MRF are usually composed by a polymeric matrix with small crystals of some magnetic material embedded on it, for example, magnetite. When the particles are superparamagnetic, the quality of the magnetic response is improved because the imanation curve has neither hysteresis nor remanence.

Another point of view for classifying these suspensions is the rheological perspective. By rheology, we name the discipline which study deformations and flowing of materials when some stress is applied. In some ranges, it is possible to consider the magnetic colloids as Newtonian fluids because, when an external magnetic field is applied, the stress is proportional to the velocity of the deformation. On a more global perspective, these fluids can be immersed on the category of complex fluids (Larson (1999)) and are studied as complex systems (Science. (1999)).

Now we are going to briefly provide some details about magnetic interactions: magnetic dipolar interaction, interaction between chains and irreversible aggregation.

2.1.1 Magnetic dipolar interaction

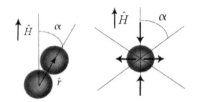

Fig. 1. Left: Two magnetic particles under a magnetic field $\vec{H}$. The angle between the field direction and the line that join the centres of the particles is named as α. Right: The attraction cone of a magnetic particle. Top and bottom zones are magnetically attractive, while regions on the left and on the right have repulsive behaviour.

As it has been said before, the main interest of MRF are their properties in response to external magnetic fields. These properties can be optical (birefringence (Bacri et al. (1993)), dichroism (Melle (2002))) or magnetical or rheological. Under the action of an external magnetic field, the particles acquire a magnetic moment and the interaction between the magnetic moments generates the particles aggregation in the form of chain-like structures. More in detail, when a magnetic field $\vec{H}$ is applied, the particles in suspension acquire a dipolar moment:

$$\vec{m} = \frac{4\pi a^3}{3} \vec{M} \tag{1}$$

where $\vec{M} = \chi \vec{H}$ and a are respectively the particle's imanation and radius, whereas χ is the magnetic susceptibility of the particle.

The most simple way for analysing the magnetic interaction between magnetic particles is through the dipolar approximation. Therefore, the interaction energy between two magnetic dipoles $\vec{m}_i$ and $\vec{m}_j$ is:

$$U_{ij}^d = \frac{\mu_0 \mu_s}{4\pi r^3} \left[(\vec{m}_i \cdot \vec{m}_j) - 3(\vec{m}_i \cdot \hat{r})(\vec{m}_j \cdot \hat{r}) \right] \tag{2}$$

where $\vec{r}_i$ is the position vector of the particle i, $\vec{r} = \vec{r}_j - \vec{r}_i$ joins the centre of both particles and $\hat{r} = \vec{r}/r$ is its unitary vector.

Then, we can obtain the force generated by $\vec{m}_i$ under $\vec{m}_j$ as:

$$\vec{F}_{ij}^d = \frac{3\mu_0\mu_s}{4\pi r^4} \left\{ \left[(\vec{m}_i \cdot \vec{m}_j) - 5(\vec{m}_i \cdot \hat{r})(\vec{m}_j \cdot \hat{r}) \right] \hat{r} + (\vec{m}_j \cdot \hat{r})\vec{m}_i + (\vec{m}_i \cdot \hat{r})\vec{m}_j \right\} \tag{3}$$

If both particles have identical magnetic properties and knowing that the dipole moment aligns with the field, we obtain the following two expressions for potential energy and force:

$$U_{ij}^d = \frac{\mu_0\mu_s m^2}{4\pi r^3}(1 - 3\cos^2\alpha) \tag{4}$$

$$\vec{F}_{ij}^d = \frac{3\mu_0\mu_s m^2}{4\pi r^4} \left[(1 - 3\cos^2\alpha)\hat{r} - \sin(2\alpha)\hat{\alpha} \right] \tag{5}$$

where α is the angle between the direction of the magnetic field $\hat{H}$, and the direction set by $\hat{r}$ and where $\hat{\alpha}$ is its unitary vector.

From the above equations, it follows that the radial component of the magnetic force is attractive when $\alpha < \alpha_c$ and repulsive when $\alpha > \alpha_c$, where $\alpha_c = \arccos \frac{1}{\sqrt{3}} \simeq 55°$, so that the dipolar interaction defines an hourglass-shaped region of attraction-repulsion in the complementary region (see Fig.1). In addition, the angular component of the dipolar interaction always tends to align the particles in the direction of the applied magnetic field. Thus, the result of this interaction will be an aggregation of particles in linear structures oriented in the direction of $\hat{H}$.

The situation depicted here is very simplified, especially from the viewpoint of magnetic interaction itself. In the above, we have omitted any deviations from this ideal behaviour, such as multipole interactions or local field (Martin & Anderson (1996)). Multipolar interactions can become important when $\mu_p/\mu_s \gg 1$. The local field correction due to the magnetic particles themselves generate magnetic fields that act on other particles, increasing the magnetic interaction. For example, when the magnetic susceptibility is approximately $\chi \sim 1$, this interaction tends to increase the angle of the cone of attraction from 55° to about 58° and also the attractive radial force in a 25% and the azimuth in a 5% (Melle (2002)).

One type of fluid, called electro-rheological (ER fluids) is the electrical analogue of MRF. This type of fluid is very common in the study of kinematics of aggregation. Basically, the ER fluids consist of suspensions of dielectric particles of sizes on the order of micrometers (up to hundreds of microns) in conductive liquids. This type of fluid has some substantial differences with MRF, especially in view of the ease of use. The development of devices using electric fields is more complicated, requiring high power voltage; in addition, ER fluids have many more problems with surface charges than MRF, which must be minimized as much as possible in aggregation studies. However, basic physics, described above, are very similar in both systems, due to similarities between the magnetic and electrical dipolar interaction.

2.1.2 Magnetic interaction between chains

Chains of magnetic particles, once formed, interact with other chains in the fluid and with single particles. In fact, the chains may laterally coalesce to form thicker strings (sometimes called columns). This interaction is very important, especially when the concentration of particles in suspension is high. The first works that studied the interaction between chains of particles come from the earliest studies of external field-induced aggregation (Fermigier & Gast (1992); Fraden et al. (1989))

Basically, the aggregation process has two stages: first, the chains are formed on the basis of the aggregation of free particles, after that, more complex structures are formed when chains aggregate by lateral interaction. When the applied field is high and the concentration of particles in the fluid is low, the interactions between the chains are of short range. Under this situation, there are two regions of interaction between the chains depending on the lateral distance between them: when the distance between two strings is greater than two diameters of the particle, the force is repulsive; if the distance is lower, the resultant force is attractive, provided that one of the chains is moved from the other a distance equal to one particle's radius in the direction of external field (Furst & Gast (2000)). In this type of interactions, the temperature fluctuations and the defects in the chains morphology are particularly important. Indeed, variations on these two aspects generate different types of theoretical models for the interaction between chains. The model that takes into account the thermal fluctuations in the structure of the chain for electro-rheological fluids is called *HT* (Halsey & Toor (1990)), and was subsequently extended to a modified HT model (MHT) (Martin et al. (1992)) to include dependence on field strength. The latter model shows that only lateral interaction occurs between the chains when the characteristic time associated with their thermal relaxation is greater than the characteristic time of lateral assembling between them. Possible defects in the chains can vary the lateral interaction, mainly through perturbations in the local field.

2.1.3 Irreversible aggregation

The irreversible aggregation of colloidal particles is a phenomenon of fundamental importance in colloid science and its applications. Basically, there are two basic scenarios of irreversible colloidal aggregation. The first, exemplified by the model of Witten & Sander (1981), is often referred to as Diffusion-Limited Aggregation (DLA). In this model, the particles diffuse without interaction between them, so that aggregation occurs when they collide with the central cluster. The second scenario is when there is a potential barrier between the particles and the aggregate, so that aggregation is determined by the rate at which the particles manage to overcome this barrier. The second model is called Colloid Reaction-Limited Aggregation (RLCA). These two processes have been observed experimentally in colloidal science (Lin et al. (1989); Tirado-Miranda (2001)).

These aggregation processes are often referred as fractal growth (Vicsek (1992)) and the aggregates formed in each process are characterized by a concrete fractal dimension. For example, in DLA we have aggregates with fractal dimension $D_f \sim 1.7$, while RLCA provides $D_f \sim 2.1$. A very important property of these systems is precisely that its basic physics is independent of the chemical peculiarities of each system colloidal i.e., these systems have universal aggregation. Lin et al. (1989) showed the universality of the irreversible aggregation systems performing light scattering experiments with different types of colloidal particles and changing the electrostatic forces in order to study the RLCA and DLA regimes in a differentiated way. They obtained, for example, that the effective diffusion coefficient (Eq.28) did not depend on the type of particle or colloid, but whether the process aggregation was DLA or RLCA.

The DLA model was generalized independently by Meakin (1983) and Kolb et al. (1983), allowing not only the diffusion of particles, but also of the clusters. In this model, named Cluster-Cluster Aggregation (CCA), the clusters can be added by diffusion with other clusters or single particles. Within these systems, if the particles are linked in a first touch, we obtain the DCLA model. The theoretical way to study these systems is to use the theory of von Smoluchowski (von Smoluchowski (1917)) for cluster-cluster aggregation among Monte Carlo simulations (Vicsek (1992)). This theory considers that the aggregation kinetics of a system of N particles, initially separated and identical, aggregate; and these clusters join themselves to form larger objects. This process is studied through the distribution of cluster sizes $n_s(t)$ which can be defined as the number of aggregates of size s per unit of volume in the system at a time t. Then, the temporal evolution is given by the following set of equations:

$$\frac{dn_s(t)}{dt} = \frac{1}{2} \sum_{i+j=s} K_{ij} n_i n_j - n_s \sum_{j=1} K_{sj} n_j, \tag{6}$$

where the kernel K_{ij} represents the rate at which the clusters of size i and j are joined to form a cluster of size $s = i + j$. All details of the physical system are contained in the kernel K_{ij}, so that, for example, in the DLA model, the kernel is proportional to the product of the cross-section of the cluster and the diffusion coefficient. Eq.6 has certain limitations because only allows binary aggregation processes, so it is just applied to processes with very low concentration of particles.

A scaling relationship for the cluster size distribution function in the DCLA model was introduced by Vicsek & Family (1984) to describe the results of Monte Carlo simulations. This scaling relationship can be written as:

$$n_s \sim s^{-2} g\left(s/S(t)\right) \tag{7}$$

where $S(t)$ is the average cluster size of the aggregates:

$$S(t) \equiv \frac{\sum_s s^2 n_s(t)}{\sum_s s n_s(t)} \tag{8}$$

and where the function $g(x)$ is in the form:

$$g(x) \begin{cases} \sim x^\Delta & \text{if } x \ll 1 \\ \ll 1 & \text{if } x \gg 1 \end{cases}$$

One consequence of the scaling 7 is that a temporal power law for the average cluster size can be deduced:

$$S(t) \sim t^z \tag{9}$$

Calculating experimentally the average cluster size along time, we can obtain the kinetic exponent z. Similarly to $S(t)$ is possible to define an average length in number of aggregates $l(t)$:

$$l(t) \equiv \frac{\sum_s s\, n_s(t)}{\sum_s n_s(t)} = \frac{1}{N(t)} \sum_s s\, n_s(t) = \frac{N_p}{N(t)} \tag{10}$$

where $N(t) = \sum n_s(t)$ is the total number of cluster in the system at time t and $N_p = \sum s\, n_s(t)$ is the total number of particles. Then, it is expected that N had a power law form with exponent z':

$$N(t) \sim t^{-z'} \tag{11}$$

$$l(t) \sim t^{z'} \tag{12}$$

Hydrodynamics on Charged Superparamagnetic Microparticles in Water Suspension: Effects of Low-Confinement Conditions and Electrostatics Interactions

59

2.2 Brownian movement and microrheology

Robert Brown[1] (1773-1858) discovered the phenomena that was denoted with his name in 1827, when he studied the movement of pollen in water. The explanation of Albert Einstein in 1905 includes the named Stokes-Einstein relationship for the diffusion coefficient of a particle of radius a immersed in a fluid of viscosity η at temperature T:

$$D = \frac{k_B T}{6\pi a \eta} \tag{13}$$

where k_B is the Boltzmann constant. This equation can be generalized for an object (an aggregate) formed by a number of particles N:

$$D = \frac{k_B T}{6\pi \eta R_g}$$

where R_g is the radius of gyration defined as $R_g(N) = \sqrt{1/N \sum_{i=1}^{N} r_i^2}$, where r_i is the distance between the i particle to the centre of mass of the cluster. If $R_g = a$, we recover the Stokes-Einstein expression.

Let's see how to calculate the diffusion coefficient D from the observation of individual particles moving in the fluid. The diffusion equation says that:

$$\frac{\partial \rho}{\partial t} = D\nabla^2 \rho$$

where ρ is here the probability density function of a particle that spreads a distance Δr at time t. This equation has as a solution:

$$\rho(\Delta r, t) = \frac{1}{(4\pi Dt)^{3/2}} e^{-\Delta r^2 / 4Dt} \tag{14}$$

If the Brownian particle moves a distance Δr in the medium on which is immersed after a time δt, then the mean square displacement (MSD) weighted with the probability function given by Eq.14 is given by:

$$\left\langle (\Delta r)^2 \right\rangle = \left\langle |r(t + \delta t) - r(t)|^2 \right\rangle = 6Dt \tag{15}$$

The diffusion coefficient can be obtained by 15 and observing the displacement Δr of the particle for a fixed δt. In two dimensions, the equations 14 and 15 are:

$$\rho(\Delta r, t) = \frac{1}{(4\pi Dt)} e^{-\Delta r^2 / 4Dt} \tag{16}$$

$$\left\langle |r(t + \delta t) - r(t)|^2 \right\rangle = 4Dt \tag{17}$$

The equations 13 and 15 are the basis for the development of a experimental technique known as microrheology (Mason & Weitz (1995)). This technique consists of measuring viscosity and other mechanical quantities in a fluid by monitoring, using video-microscopy, the movement

---

[1] Literally: *While examining the form of these particles immersed in water, I observed many of them very evidently in motion [..]. These motions were such as to satisfy me, after frequently repeated observation, that they arose neither from currents in the fluid, nor from its gradual evaporation, but belonged to the particle itself.* (Edinburgh New Philosophical Journal, Vol. 5, April to September, 1828, pp. 358-371)

of micro-nano particles (regardless their poralization). Thus, it is possible to obtain the viscosity of the medium simply by studying the displacement of the particle in the fluid. The microrheology has been widely used since the late nineties of last century (Waigh (2005)). Due to microrheology needs and for the sake of the analysis of the thermal fluctuation spectrum of probe spheres in suspension, the generalized Stokes-Einstein equation (Mason & Weitz (1995)) was developed. This expression is similar to Eq.13, but introducing Laplace transformed quantities:

$$\tilde{D}(s) = \frac{k_B T}{6\pi a s \, \tilde{\eta}_s} \tag{18}$$

where s is the Laplace frequency, and $\tilde{\eta}_s$ and $\tilde{D}(s)$ are the Laplace transformed viscosity and diffusion coefficient. The dynamics of the Brownian particles can be very different depending on the mechanical properties of the fluid. This equation is the base for the rheological study, by obtaining its viscoelastic moduli (Mason (2000)), of the complex fluid in which the particles are immersed.

If we only track the random motion of colloidal spheres moving freely in the fluid, we are talking of "passive" microrheology, but there are variations on this technique named "active" microrheology, for example, using optical tweezers (Grier (2003)). This technique allows to study the response of colloidal particles in viscoelastic fluids and the structure of fluids in the micro-nanometer scales (Furst (2005)), measure viscoelastic properties of biopolymers (like DNA) and the cell membrane (Verdier (2003)). Other useful methodologies are the two-particles microrheology (Crocker et al. (2000)) which allows to accurately measure rheological properties of complex materials, the use of rotating chains following an external rotating magnetic field (Wilhelm, Browaeys, Ponton & Bacri (2003); Wilhelm, Gazeau & Bacri (2003)) or magnetic bead microrheometry (Keller et al. (2001)).

2.3 Hydrodynamics

When we are talking about hydrodynamics in a colloidal suspension of particles we need to introduce the Reynolds number, Re, defined as:

$$\text{Re} \equiv \frac{\rho_r \, v \, a}{\eta} \tag{19}$$

where ρ_r is the relative density, a is the particle radius, v is the velocity of the particle in the fluid which has a viscosity η. This number reflects the relation between the inertial forces and the viscous friction. If we are in a situation of low Reynolds number dynamics, as it usually happens in the physical situation here studied, the inertial terms in the Newton equations can be neglected, and $m\ddot{x} \cong 0$.

However, even in the case of low Reynolds number, the diffusion coefficient of particles in a colloidal system may have certain deviations from the expressions explained above. The diffusion coefficient can vary due to hydrodynamic interactions between particles, the morphology of the clusters, or because of the enclosure containing the suspension. When a particle moves near a "wall", the change in the Brownian dynamics of the particle is remarkable. The effective diffusion coefficient then varies with the distance of the particle from the wall (Russel et al. (1989)), the closer is the particle to the wall, the lower the diffusion coefficient. The interest of the modification on Brownian dynamics in confinement situations is quite large, for example to understand how particles migrate in porous media, how the macromolecules spread in membranes, or how cells interact with surfaces.

Hydrodynamics on Charged Superparamagnetic Microparticles in Water Suspension: Effects of Low-Confinement Conditions and Electrostatics Interactions

61

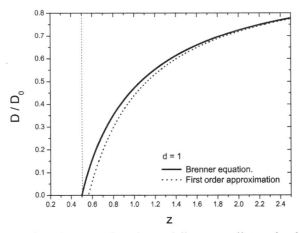

Fig. 2. Comparative analysis between the relative diffusion coefficient for the Brenner equation (Eq.20) and the first order approximation (Eq.21), as a function of the distance to the wall z for a particle of diameter 1 (z-unit are in divided by the diameter of the particle). These two expressions are practically equal when $z \geq 1.5$.

2.3.1 Particle-wall interaction

When a particle diffuses near a wall, thanks to the linearity of Stokes equations, the diffusion coefficient can be separated into two components, one parallel to the wall $D_{\parallel}$ and the other perpendicular $D_{\perp}$. In the literature, several studies in this regard can be found (Crocker (1997); Lin et al. (2000); Russel et al. (1989)). One particularly important is the study of Faucheux & Libchaber (1994) where measurements of particles confined between two walls are reported. This work provides a table with the diffusion coefficients obtained (theoretical and experimental) for different samples (different radius and particles) and different distances from the wall, from 1 to 12 μm. For example, for a particle diameter 2.5 μm, a distance of 1.3 μm from the wall and with a density 2.1 times that of water, a diffusion coefficient $D/D_0 = 0.32$ is obtained, where D_0 is the diffusion coefficient given by Eq.13.

There are no closed analytical solutions for this type of problem, with the exception of that obtained for a sphere moving near a flat wall in the direction perpendicular to it (Brenner (1961)):

$$\frac{D_{\perp}(z)}{D_0} = \left\{ \frac{4}{3} \sinh \alpha \sum_{n=1}^{\infty} \frac{n(n+1)}{(2n-1)(2n+3)} \left[\frac{2\sinh[(2n+1)\alpha] + (2n+1)\sinh[2\alpha]}{4\sinh^2[(n+1/2)\alpha] - (2n+1)^2 \sinh^2[\alpha]} - 1 \right] \right\}^{-1}$$
(20)

where $\alpha \equiv \text{arccosh}\,(z/a)$ and a is the radius particle and z is the distance between the centre of the particle and the wall.

Theoretical calculations in this regard are generally based on the methods of reflections, which involves splitting the hydrodynamic interaction between the wall and the particle in a linear superposition of interactions of increasing order. Using this method, it is possible to obtain an iterative solution for this problem in power series of (a/z). In the case of the perpendicular direction it is found:

$$\frac{D_{\perp}(z)}{D_0} \cong 1 - \frac{9}{8}\left(\frac{a}{z}\right) + O\left(\frac{a}{z}\right)^3$$
(21)

In the Fig.2 a comparison between the exact equation 20 and this first order expression 21 is plotted. These two expressions provide similar results when $z \geq 1.5$.

In the case of the parallel direction to the wall we have the following approximation :

$$\frac{D_{||}(z)}{D_0} \cong 1 - \frac{9}{16}\frac{a}{z} + \frac{1}{8}\frac{a^3}{z^3} - \frac{45}{256}\frac{a^4}{z^4} - \frac{1}{16}\frac{a^5}{z^5} + \dots \tag{22}$$

which is commonly used in their first order:

$$\frac{D_{||}(z)}{D_0} \cong 1 - \frac{9}{16}\left(\frac{a}{z}\right) + O\left(\frac{a}{z}\right)^3 \tag{23}$$

If we are thinking about one particle between two close walls, Dufresne et al. (2001) showed how it is possible to deduce, using the Stokeslet method (Liron & Mochon (1976)), a very complicated closed expression for the diffusion coefficients when $a \ll h$, being h the distance between the two walls. However, the method of reflections gives approximated theoretical expressions. Basically, there are three approximations that provide good results and which are different because of small modifications in the drag force. The first of these methods is the Linear Superposition Approximation (LSA) where the drag force over the sphere is chosen as the sum of the force that makes all the free fluid over the sphere. A second method is the Coherent Superposition Approximation (CSA) whose modification proposed by Bensech & Yiacoumi (2003) was named as Modified Coherent Superposition Approximation (MCSA) and gives the following expression:

$$\frac{D(z)}{D_0} = \left\{ 1 + [C(z) - 1] + [C(h - z) - 1] + \sum_{n=1}^{\infty}(-1)^n \frac{nh - z - a}{nh - z}[C(nh + z) - 1] \right.$$
$$\left. + \sum_{n=1}^{\infty}(-1)^n \frac{(n-1)h + z - a}{(n-1)h + z}[C((n+1)h - z) - 1] \right\}^{-1} \tag{24}$$

where the function $C(z)$ is the inverse of the normalized diffusion coefficient $(D_0/D(z))$ in the only one wall situation.

Another interesting physical configuration is the hydrodynamic coupling of two Brownian spheres near to a wall. Dufresne et al. (2000) showed that the collective diffusion coefficients in the directions parallel and perpendicular to the surface are related by a hydrodynamical coupling because of the fact that the surrounded fluid moved by one of the particles affects the other. This wall-induced effect may have an influence in the origin of some anomalous effects in experiments of confined microparticles in suspension.

2.3.2 Particle-particle interaction

Another effect of considerable importance, or at least, that we must take into account, is the hydrodynamic interaction between two particles. This effect is quantified by the parameter $\rho = r/a$ where r is the radial distance between the centres of the particles and a is their radius. Crocker (1997) showed how the modification of the diffusion coefficient due to the mutual hydrodynamic interaction between the two particles varies in the directions parallel or perpendicular to the line joining the centres of mass. Finally, they obtained that the predominant effect is the one that occurs in the radial direction and which is given by:

$$\frac{D}{D_0} \cong -\frac{15}{4\rho^4} \tag{25}$$

The effect in the perpendicular direction is much lower and negligible ($O(\rho^{-6})$).

2.3.3 Anisotropic friction

When the aggregates are formed in the suspensions, their way of spreading in the fluid is expected to change. By analogy with the Stokes-Einstein equation, in which the diffusion coefficient depends on the inverse of particle diameter ($D \sim a^{-1}$), Miyazima et al. (1987) suggested that the diffusion coefficient depends on the inverse cluster size s in the form $D(s) \sim s^{\gamma}$, where γ is the coefficient that marks the degree of homogeneity of the kernel on the Smoluchowski equation (Eq. 6). The result for the diffusion coefficient $\gamma = -1$ is considered to be strictly valid for spherical particles that not interact hydrodynamically among them. However, in the case of an anisotropic system, as is the case of chain aggregates, the diffusion coefficient varies due to the hydrodynamic interaction in the direction parallel and perpendicular to the axis of the chain, as follows (Doi & Edwards (1986)):

$$D_{\parallel} = \frac{k_B T}{2\pi\eta a} \frac{\ln s}{s} \tag{26}$$

$$D_{\perp} = D_{\parallel}/2 \tag{27}$$

This result is based assuming point particles, but similar expressions are obtained by modelling the aggregates in the form of cylinders of length L and diameter $d = 2a$. Tirado & García (1979; 1980) provide diffusion coefficients for this objects in the directions perpendicular, parallel and rotational to the axis of the chains ($D_{\parallel}, D_{\perp}, D_r$). By using mesaurements of Dynamic Light Scattering (DLS), an effective diffusion coefficient, D_{eff}, of the aggregates can be extracted (Koppel (1972)). This effective coefficient is related to the others mentioned above by means of the relationship:

$$D_{\text{eff}} = D_{\perp} + \frac{L^2}{12} D_r \tag{28}$$

which is correct if $qL >> 1$ where q is the scattering wave vector defined as: $q = 4\pi/\lambda_l \sin(\theta/2)$, λ_l is the wave length of the laser over the suspension and θ is the scattering angle.

2.3.4 Cluster sedimentation

A particularly important effect is the sedimentation of the clusters or aggregates. It is essential, when a colloidal system is studied, determine the position of the aggregates from the wall, as well as knowing what the deposition rate by gravity is and when the equilibrium in a given layer of fluid is reached. The velocity v_c experienced by a cluster composed of N identical spherical particles of radius a and mass M falling by gravity in a fluid without the presence of walls is (González et al. (2004)):

$$v_c = \frac{MgN}{\gamma_0}\left(1 - \frac{\rho}{\rho_p}\right) = \frac{MgDN}{k_B T}\left(1 - \frac{\rho}{\rho_p}\right)$$

where g is the value of the gravity acceleration, ρ is the fluid density, ρ_p is the density of the particles, γ_0 is t the drag coefficient and D the diffusion coefficient. If we have only one spheric particle, the last equation yields the classic result for the sedimentation velocity:

$$v_p = \frac{2a^2 g \Delta\rho}{9\eta}$$

with $\Delta\rho = \rho_p - \rho$. We can define the Péclet number as the ratio between the sedimentation time t_s and diffusion t_d using a fixed distance, for instance, $2a$:

$$P_e \equiv \frac{t_d}{t_s} = \frac{Mga}{k_B T}\left(1 - \frac{\rho}{\rho_p}\right) = \frac{4\pi a^4 g \Delta\rho}{3k_B T} \qquad (29)$$

Then, the vertical distance travelled by gravity for a cluster in a time equal to that a particle spread a distance equal to the diameter of the particle d is $d_c = v_c t_d = P_e N d$.

The above expressions are satisfied when sedimentation occurs in an unconfined fluid. If there is a bottom wall, then it provides a spatial distribution of particles ρ which depends on the relative height with respect to the bottom wall. If the system is in an equilibrium state and with low concentration of particles, we can use the Boltzmann density profile, which measures the balance on the thermal forces and gravity:

$$\ln\rho(z) \propto -\frac{z}{L_G} \qquad (30)$$

where $L_G \sim k_B T / Mg$. As mentioned, this density profile is valid when the interactions between the colloidal particles are neglected. However, experimental situations can be much more complicated, resulting in deviations from this profile, so theoretical research is still in development about this question (Chen & Ma (2006); Schmidt et al. (2004)). In fact, it has been discovered experimentally that the influence of the electric charge of silica nanoparticles in a suspension of ethanol may drastically change the shape of the density profile (Rasa & Philipse (2004)). We will here assume the expression 30 to be correct, so that the average height z_m of a particle of radius a, between two walls separated by a distance h, can be determined by the Boltzmann profile as Faucheux & Libchaber (1994) showed:

$$P_B(z) = \frac{1}{L}\left(\frac{e^{-z/L}}{e^{-a/L} - e^{(a-h)/L}}\right) \qquad (31)$$

where z is the position of the particle between the two walls, where the bottom wall is at $z = 0$ and the top is located at $z = h$, L is the characteristic Boltzmann length defined as $L \equiv k_B T (g\Delta M)^{-1}$ where $\Delta M \equiv (4/3)\pi a^3(\rho_p - \rho)$.

Therefore, the mean distance z_m can be calculated:

$$z_m = \int_a^{h-a} zP_B(z)dz = \qquad (32)$$
$$= \frac{e^{-a/L}[aL + L^2] - e^{(a-h)/L}[(h-a)L + L^2]}{L[e^{-a/L} - e^{(a-h)/L}]}$$

With that expression and the equations for the diffusion coefficient near a wall (Eqs. 20 to 25) we can estimate the effective diffusion coefficient of a sedimented particle. However, when we have a set of particles, clusters or aggregates near the walls of the enclosure, the evaluation of hydrodynamic effects on the diffusion coefficient and their dynamics is not an easy problem to evaluate theoretically or experimentally. In fact, this problem is very topical, for example, focused on polymer science (Hernández-Ortiz et al. (2006)) or more specifically, in the case of biopolymers, such as DNA strands, moving by low flows in confined enclosures (Jendrejack et al. (2003)). Kutthe (2003) performed Stokestian dynamics simulations (SD) of chains, clusters and aggregates in various situations in which hydrodynamic interactions are not negligible. Specifically, they calculated the friction coefficient γ_N depending on N number of particles) for linear chains located at a distance z of the wall and applying a transverse velocity $V_x = 0.08$ diameters per second. The friction coefficient γ_N, to reach a velocity V_x in the transverse direction was obtained as:

$$\gamma_N = \frac{F_x}{3\pi\eta dV_x}$$

where F_x is the force over the chain and d the diameter of the particle. Then, they obtain that, far away from the wall, $\gamma_{30} \sim 6$ for a chain formed by 30 particles. But, near enough from the wall, the friction coefficient grows to a value $\gamma_{30} \sim 200$. Recently, Paddinga & Briels (2010) showed simulation results for translational and rotational friction components of a colloidal rod near to a planar hard wall. They obtained a enhancement friction tensor components because of the hydrodynamic interactions between the rod and the wall.
In any case, when we are thinking on one spherical Brownian particle, we can estimate the diffusion coefficient using the Boltzmann profile by calculating the mean position of the particle using Eq.32. Then, if we can calculate the experimental diffusion coefficient when sedimentation affects to the particles, we can employ the following expression Domínguez-García, Pastor, Melle & Rubio (2009); Faucheux & Libchaber (1994)):

$$D_{\parallel}^{\delta} = \int_0^L P_B(z) \left[\int_{z-\delta(z,\eta)}^{z+\delta(z,\eta)} D_{\parallel}(z',\eta) \frac{P_B(z')}{N_B(z',\eta)} dz' \right] dz$$

where $P_B(z)$ is the Boltzmann probability distribution, $N_B(z)$ is the normalization of that function, $D_{\parallel}(z',\eta)$ is the corrected diffusion coefficient of the particle for the motion parallel to the wall. This expression introduces a correction because of the vertical movement: during each time window of span τ, the particle typically explores a region of size 2δ with $\delta(z,\eta) = \frac{1}{2}\sqrt{2\tau D_{\perp}(z,\eta)}$, where $D_{\perp}$ is the diffusion coefficient for the motion normal to the wall. The height of the particle from the bottom, z, is calculated by assuming the Boltzmann probability distribution.

2.4 Electrostatics

In a colloidal system, there are usually present not only external forces or hydrodynamic interaction of particles with the fluid, but also electrostatic interactions of various kinds. Moreover, as we shall see, many of the commercial micro-particles have carboxylic groups ($-COOH$) to facilitate their possible use, for example, in biological applications. These groups provide for electrolytic dissociation, a negative charge on the particle surface, so that we can see their migration under a constant and uniaxial electric field using the technique of electrophoresis. Therefore, these groups generate an electrostatic interaction between the particles.

2.4.1 DLVO theory

DLVO theory (Derjaguin & Landau (1941); Verwey & Overbeek (1948)) is the commonly used classical theory to explain the phenomena of aggregation and coagulation in colloidal particle systems without external fields applied. Roughly speaking, the theory considers that the colloidal particles are subject to two types of electrical forces: repulsive electrostatic forces due to same-sign charged particles and, on the other hand, Van der Waals forces which are of attractive nature and appear due to the interaction between the molecules that form the colloid. According to the intensity relative to each other, the particles will aggregate or repel.

Thus, the method to control the aggregation is to vary the ionic strength of medium, i.e., the pH. In most applications in colloids, it is enormously important to control aggregation of particles, for example, for purification treatments of water.

The situation around a negatively charged colloidal particle is approximately described by the double layer model. This model is used to display the ionic atmosphere in the vicinity of the charged colloid and explain how the repulsive electrical forces act. Around the particle, the negative charge forms a rigid layer of positive ions from the fluid, usually called Stern layer. This layer is surrounded by the diffuse layer that is formed by positive ions seeking to approach the colloidal particle and that are rejected by the Stern layer. In the diffuse layer there is a deficit of negative ions and its concentration increases as we left the colloidal particle. Therefore, the diffuse layer can be viewed as a positively charged atmosphere surrounding the colloid.

The two layers, the Stern layer and diffuse layer, form the so-called double layer. Therefore, the negative particle and its atmosphere produce a positive electrical potential associated with the solution. The potential has its maximum value on the surface of the particle and gradually decreases along the diffuse layer. The value of the potential that brings together the Stern layer and the diffuse layer is known as the Zeta potential, whose interest mainly lies in the fact that it can be measured. This Zeta potential measurement, is commonly referred as ζ and measured in mV. The Zeta potential is usually measured using the Laser Doppler Velocimeter technique. This device applies an electric field of known intensity of the suspension, while this is illuminated with a laser beam. The device measures the rate at which particles move so that the Zeta potential, ζ, can be calculated by several equations that relate the Zeta potential electrophoretic mobility, μ_e.

In a general way, it is possible to use the following expression, known as the Hückel equation:

$$\mu_e = \frac{2}{3}\frac{\varepsilon \zeta}{\eta} f(\kappa a) \tag{33}$$

where ε is the dielectric constant of the medium, η its viscosity, a the radius of the particle and where $1/\kappa$ is the width of the double layer, known as the Debye screening length and where $f(\kappa a)$ is the named Henry function. In the case of $1 < \kappa a < 100$, the Zeta potentials can be calculated by means of some analytic expression of the Henry function (Otterstedt & Brandreth. (1998)). Summarising, the higher is the Zeta potential, the more intense will be the Coulombian repulsion between the particles and the lower will be the influence of the Van der Waals force in the colloid.

The Van der Walls potential, which can provide a strong attractive interaction, is usually neglected because its influence is limited to very short surface-to-surface distances in the order of 1 nm. Therefore, the DLVO electrostatic potential between two particles located a radial distance r one from the other is usually given by the classical expression:

$$U(r) = \frac{(Z^*e)^2}{\varepsilon}\frac{\exp(2a\kappa)}{(1+a\kappa)^2}\frac{\exp(-\kappa r)}{r} \tag{34}$$

where Z^* is the effective charge of the particles and $\sigma_{\text{eff}} = Z^*e/4\pi a^2$ is their density of effective charge. Therefore, in this theory, two spherical like-charged colloidal particles suffered a purely electrostatic repulsion between them. The colloidal particle can have carboxylic groups ($COOH$) attached to their surfaces, creating a layer of negative charge of length δ in the order of nanometers surrounding the colloidal particles (Shen et al. (2001)).

The presence of this layer modifies the equation of the double-layer potential (Reiner & Radke 1993); Shen et al. (2001)):

$$U_{dl}(s) = 2\pi\varepsilon(\psi)^2 \frac{2}{2 + s'/a} \exp(-\kappa s')$$ (35)

where $s' = s - 2\delta$.

2.4.2 Ornstein-Zernike equation

For calculating the electrostatic potential in a colloidal suspension, we can use the following methodology. This approach involves using the radial distribution function of the particles, $g(r)$, knowing that it is related with the interaction energy of two particles in the limit of infinite dilution by means of the Boltzmann distribution:

$$\lim_{n \to 0} g(r) = e^{-\beta U(r)}$$ (36)

where n is the particle density and $\beta \equiv 1/k_B T$. However, for finite concentrations, $g(r)$ is influenced by the proximity between particles, so we can calculate the mean force potential, $w(r)$:

$$w(r) = -\frac{1}{\beta} \ln g(r)$$ (37)

But we do not know the relation between $w(r)$ and $U(r)$. Here, is usually defined a total correlation function $h(r) \equiv g(r) - 1$ and is used the Ornstein-Zernike (O-Z) equation for two particles in a two-dimensional fluid:

$$h(r) = c(r) + n \int c(r')h(|\mathbf{r}' - \mathbf{r}|)dr'$$ (38)

The $c(r)$ function is the direct correlation function between two particles. Now, it is necessary to close the integral equation by linking $h(r)$, $c(r)$ and $U(r)$. For that, one of the following assumptions is employed:

$$c(r) = \begin{cases} -\beta U(r) & \text{MSA} \\ -\beta U(r) + h(r) - \ln g(r) & \text{HNC} \\ (1 - e^{\beta U})(1 + h(r)) & \text{PY} \end{cases}$$ (39)

named Mean Spherical Approximation (MSA), Hypernetted Chain (HNC) and Percus-Yevick (PY).

In the case of video-microscopy experiments, a more practical methodology is explained by Behrens & Grier (2001b) for obtaining the electrostatic potential. More explicitly, with the PY approximation we have:

$$U(r) = w(r) + \frac{n}{\beta} I(r) = -\frac{1}{\beta} [\ln g(r) - nI(r)],$$ (40)

and with the HC:

$$U(r) = w(r) + \frac{1}{\beta} \ln[1 + nI(r)] = -\frac{1}{\beta} \left[\ln \left(\frac{g(r)}{1 + nI(r)} \right) \right],$$ (41)

In both cases, $I(r)$ is the convolution integral defined as:

$$I(r) = \int [g(r') - 1 - nI(r')] [g(|\mathbf{r}' - \mathbf{r}|) - 1] d^2r',$$ (42)

which can be calculated numerically.

2.4.3 Anomalous effects

In order to understand the interactions in this kind of systems, we have to note that the standard theory of colloidal interactions, the DLVO theory, fails to explain several experimental observations. For example, an attractive interaction is observed between the particles when the electrostatic potential is obtained. This is a effect that has been previously observed in experiments on suspensions of confined equally-charged microspheres (Behrens & Grier (2001a;b); Grier & Han (2004); Han & Grier (2003); Larsen & Grier (1997)). Grier and colleagues listed several experimental observations using suspensions of charged polystyrene particles with diameters around 0.65 microns at low ionic strength and strong spatial confinement. They note that such effects appear when a wall of glass or quartz is near the particles. Studying the $g(r)$ function and its relation to the interaction potential, given by expression 36, they showed the appearance of a minimum on the potential located at $z = 2.5$ microns of the wall and a distance between centres to be $r_{min} = 3.5$ microns. This attraction cannot be a Van der Waals interaction, because for this type of particle and with separations greater than 0.1 micrometres, this force is less than $0.01 \, k_B T$ (Pailthorpe & Russel (1982)), while this attractive interaction is about $0.7 \, k_B T$.

The same group (Behrens & Grier (2001b)) extended this study using silica particle suspensions (silicon dioxide, SiO_2) of 1.58 microns in diameter, with a high density of 2.2 g/cm^3, using a cell of thickness $h = 200 \, \mu m$. In this situation, even though the particles are deposited at a distance from the bottom edge of the particle to the bottom wall equal to $s = 0.11 \, \mu m$, no minimum in the interaction energy between pairs appears, being the interaction purely repulsive, in the classical form of DLVO given by Eq.34. In that work, a methodology is also provided to estimate the Debye length of the system and the equivalent load Z^* through a study of the presence of negative charge quartz wall due to the dissociation of silanol groups in presence of water (Behrens & Grier (2001a)). However, Han & Grier (2003) observed the existence of a minimum in the potential when they use polystyrene particles of 0.65 micron and density close to water, 1.05 g/cm^3, with a separation between the walls of $h = 1.3$ microns. What is more, using silica particles from previous works, they observe a minimum separation between walls of $h = 9 \, \mu m$.

The physical explanation of this effect is not clear (Grier & Han (2004)), being the main question how to explain the influence on the separation of the two walls in the confinement cell. However, some criticism has appeared about this results. For example, about the employment of a theoretical potential with a DLVO shape. An alternative is using a Sogami-Ise (SI) potential (Tata & Ise (1998)). Moreover, Tata & Ise (2000) contend that both the DLVO theory and the SI theory are not designed for situations in confinement, so interpreting the experimental data using either of these two theories may be wrong. Controversy on the use of a DLVO-type or SI potentials appears to be resolved considering that the two configurations represent physical exclusive situations (Schmitz et al. (2003)). In fact, simulations have been performed to explore the possibility of a potential hydrodynamic coupling with the bottom wall generated by the attraction between two particles (Dufresne et al. (2000); Squires & Brenner (2000)). However, the calculated hydrodynamic effects do not seem to explain the experimental minimum on the potential (Grier & Han (2004); Han & Grier (2003)). Other authors argue that this kind of studies should be more rigorous in the analysis of errors when extracting data from the images (Savin & Doyle (2005; 2007); Savin et al. (2007)) and other authors claim that the effect on the electrostatic potential may be an artefact (Baumgart et al. (2006)) that occurs because of a incorrect extraction of the position of the particles (Gyger et al. (2008)).

)lin et al. (2007) realized that some minimums in the electrostatic potential can be eliminated
/ measuring the error on the displacement of the particles. However, this is not a double
nplication and other experimental minimums in the potential remain there. In that work,
ιe authors take into account all the proposed artefacts to date for their measurements,
emonstrating that charged glass surfaces really induce attractions between charged colloidal
)heres.   Moreover, Tata et al. (2008) claim that their observations using confocal laser
·anning of millions of charged colloidal particles establish the existence of an attractive
ehaviour in the electrostatic potential.

[oreover, other possible electrostatic variations in these systems may appear for several
asons.   For instance, the emergence of a spontaneous macroscopic electric field in
ιarged colloids (Rasa & Philipse (2004)).   Moreover, according to several studies, changes
 the fluid due to, for example, environmental pollution with atmospheric CO_2, can be
·latively easy and are not negligible at low concentrations, being able to radically change
ιe electrical properties on the fluid (Carrique & Ruiz-Reina (2009)).   Thus, interactions
·lated to colloidal stability can produce anomalous effects and significant changes in, for
<ample, sedimentation kinetics (Buzzaccaro et al. (2008)) or sedimentation-diffusion profiles
>hilipse & Koenderink (2003)).   Then, these electrostatic effects can affect the dynamics of
;gregation and influence the mobility of the particles and clusters.

. Results

·ur experimental system is formed by a MRF composed of colloidal dispersions of
ιperparamagnetic micron-sized particles in water. These particles have a radius of 485 nm
ιd a density of 1.85 g/cm^3, so they sediment to an equilibrium layer on the containing
·ll.   They are composed by a polymer (PS) with nano-grains of magnetite dispersed into
, which provide their magnetic properties.   The particles are also functionalized with
ιrboxylic groups, so they have an electrical component, therefore, they repel each other,
voiding aggregation. This effect is improved by adding sodium dodecyl sulfate (SDS) in
concentration of 1 gr/l.

he containing cell consists on two quartz windows, one of them with a cavity of 100
m.   The cell with the suspension in it is located in an experimental setup that isolate
ιermically the suspension and allows to generate a uniform external magnetic field in
ιe centre of the cell.   The particles and aggregates are observed using video-microscopy
ee details for this experimental setup on (Domínguez-García et al. (2007))).   Images of
ιe fluid are saved on the computer and then analysed for extracting the relevant data
y using our own developed software (Domínguez-García & Rubio (2009)) based on Image]
J. S. National Institutes of Health, Bethesda, Maryland, USA, http://rsb.info.nih.gov/ij/ (n.d.)).   In
ιg.3, we show an example of these microparticles and aggregates observed in our system.
he zeta potential of these particles is about −110 to −60 mV for a pH about 6 - 7. Therefore,
ιe electrical content of the particles is relatively high and it is only neglected in comparison
·ith the energy provided by the external magnetic field. However, the colloidal stability of
ιese suspensions is not being controlled and it may have an effect on the dynamics of the
usters, specially when no magnetic field is applied. In any case, as we will see, even when a
ιagnetic field is applied, it is observed a disagreement between theoretical aggregation times
ιd experimental ones.

3.1 Control parameters

We have already defined some important parameters as the Péclet number, Eq.29, an the Reynolds number Eq.19. However, in our system we need to define some externa parameters related with the concentration of particles and the intensity of the magnetic field. The concentration of volume of particles in the suspension, ϕ, is defined as the fraction c volume occupied by the spheres relative to the total volume of the suspension. In a quasi-2I video-microscopy system is useful to take into account the surface concentration ϕ_{2D}.

For measuring the influence of the magnetic interaction we used the λ parameter, defined as

$$\lambda \equiv \frac{W_m}{k_B T} = \frac{\mu_s \mu_0 m^2}{16 \pi a^3 k_B T} \tag{43}$$

as the ratio of $W_m = U_{ij}^d(r = 2a, \alpha = 0)$, i.e., the magnetic energy, and the therma fluctuations $k_B T$. Here, μ_s is the relative magnetic permeability of the solvent, μ_0 the magneti permeability of vacuum and m the magnetic moment. The parameters λ y ϕ_{2D} allow to defin a couple of characteristic lengths. First, we define a distance R_1 for which the energy of dipola interaction is equal to thermal fluctuations:

$$R_1 \equiv 2a \lambda^{1/3} \tag{44}$$

Finally, we define a mean distance between particles:

$$R_0 \equiv \sqrt{\pi} a \, \phi_{2D}^{-1/2} \tag{45}$$

The comparative between these two quantities allows to distinguish between differen aggregation regimes. When, $R_1 < R_0$, the thermal fluctuations prevail over the magneti interactions so diffusion is the main aggregation process. If $R_1 > R_0$, the aggregation of th particles occurs mainly because of the applied magnetic field.

3.2 Aggregation and disaggregation

Studies about the dynamics of the irreversible aggregation of clusters under unidirectiona constant magnetic fields have used a collection of experimental systems. For example electro-rheological fluids (Fraden et al. (1989)), magnetic holes (non-magnetic particle in a ferrofluid) (Cernak et al. (2004); Helgesen et al. (1990; 1988); Skjeltorp (1983)) and magneto-rheological fluids and magnetic particles (Bacri et al. (1993); Bossis et al (1990); Cernak (1994); Cernak et al. (1991); Fermigier & Gast (1992); Melle et al. (2001) Promislow et al. (1994)).

These studies focus their efforts in calculating the kinetic exponent z obtaining differen values ranging $z \sim 0.4 - 0.7$. The different methodologies employed can be th origin of these dispersed values. However, more recent studies (Domínguez-García et al (2007); Martínez-Pedrero et al. (2007)) suggest that this value is approximately $z \sim 0.6 - 0.7$ in accordance with experimental values reported for aggregation of dielectri colloids $z \sim 0.6$ (Fraden et al. (1989)) and with recent simulations of aggregation o superparamagnetic particles (Andreu et al. (2011)). Regarding hydrodynamics interaction Miguel & Pastor-Satorras (1999) proposed and effective expression for explaining th dispersed value of the kinetic exponent based on logarithmic corrections in the diffusio coefficient (Eqs. 26 and 27):

$$S(t) \sim (t \ln [S(t)])^{\tilde{z}}, \tag{46}$$

Hydrodynamics on Charged Superparamagnetic Microparticles in Water Suspension: Effects of Low-Confinement Conditions and Electrostatics Interactions

71

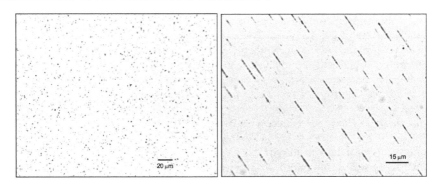

Fig. 3. Superparamagnetic microparticles observed when no external magnetic field is applied (Left) and when it is applied (Right).

where the exponent ζ is an exponent that depends on the dimensionality of the system, so if $d \geq 2, \zeta = 1/2$. Using Monte Carlo simulations they obtain that $\zeta \simeq 0.51$, and therefore that z is $z \simeq 0.61$.

In the case of our experiments, we have experimentally obtained that the z exponent in aggregation is contained in the range of $0.43 - 0.67$ (Domínguez-García et al. (2007)) with an average value of $z \sim 0.57 \pm 0.03$. These experimental values do not depend on the amplitude of the magnetic field nor on the concentration of particles, but they seem to depend on the ratio R_1/R_0, which is a sign of the more important regime of aggregation. The dependency on this ratio also appears when the morphology of the chains is studied (Domínguez-García, Melle & Rubio (2009); Domínguez-García & Rubio (2010)). Besides, the scaling behaviour given by Eq.7 is experimentally observed and checked. We have compared our experimental results with Brownian dynamics simulations based on a simple model which only included dipolar interaction between the particles, hard-sphere repulsion and Brownian diffusion, neglecting inertial terms and effects related with sedimentation or electrostatics. The results of these simulations agree with the theoretical prediction, whereas the experimental aggregation time, t_{ag}, appears to be much longer than expected (Cernak et al. (2004); Domínguez-García et al. (2007)), about three orders of magnitude of difference. The formation of dimers (two-particles aggregates) in the experiments lapses $t \sim 10^2$ seconds, but Brownian simulations show that this lapse of time is about $t \sim 0.1$ s. This last value can be easily obtained by assuming that the equation for the movement between two particles with dipolar magnetic interaction is:

$$M\ddot{r} + \gamma_0 \dot{r} + 3\mu\mu_0 m^2 r^{-4} \pi^{-1} = 0$$

where M is the mass of the particles. Because of Reynolds number (Eq.19) is very low, we neglect the inertial term on this equation. If the particles are separated a initial distance $d = R_0$ we can obtain that:

$$t_{ag} \cong \frac{32\pi\gamma_0 a^5}{15\mu_s\mu_0 m^2} \phi_{2D}^{-5/2}$$

If we express this equation in function of the λ parameter 43 and of the diffusion coefficient given by the Stokes-Einstein equation 13:

$$t_{ag} \cong \frac{2a^2}{15} \frac{1}{\lambda D} \phi_{2D}^{-5/2} \qquad (47)$$

For example, the aggregation processes for $S(t)$ in the work of Promislow et al. (1994), show an aggregation time of 200 seconds. The paramagnetic particles used in that work have a diameter of 0.6 μm and a 27% of magnetite content. Using the Stokes-Einstein expression $D = 0.86\,\mu m/s^2$ is obtained, supposing that these particles do not sediment. Using $\phi = 0.0012$ and $\lambda = 8.6$, we can obtain that $t_{ag} \sim 122$ seconds, in the order of their experimental result. In the case of our experiments, we obtain the same values using Eq.47 that using Brownian simulations.

These discrepancies may be related with hydrodynamic interactions which should affect the diffusion of the particles. From Eq.47, we see that some variation on the diffusion coefficient of the particles can modify the expected aggregation time for two particles. For testing that, we made some microrheology measurements using different types of isolated particles according to the theory of sedimentation and with the corrections on the values of the diffusion coefficient. The experimental values agree very well with the theoretical ones calculated from the expression 2.3.4 (Domínguez-García, Pastor, Melle & Rubio (2009)) but they imply a reduction on the diffusion coefficient a factor of three as a maximum, no being sufficient for explaining the discrepancy in the aggregation times.

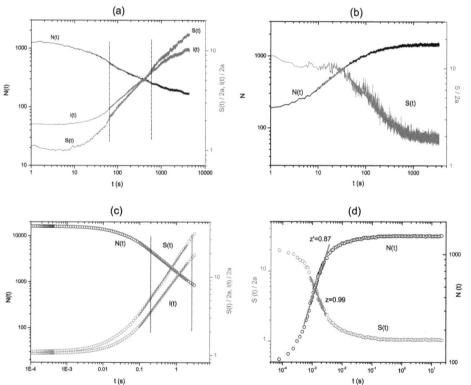

Fig. 4. Experiments of aggregation and disaggregation. The experimental process of aggregation (a) begins with $\lambda = 1718, \phi_{2D} = 0.088$ while disaggregation is shown in (b). Brownian dynamics simulations results with $\lambda = 100, \phi_{2D} = 0.03$ are shown for aggregation (c) and disaggregation (d). Data from Refs.(Domínguez-García et al. (2007; 2011))

For completing this study, we also have shown results of disaggregation, that is, the process that occurs when the external magnetic field is switched off and the clusters vanish. For this process we study the kinetics in the same way that in aggregation, by searching power laws behaviours and calculating the kinetic exponents z and z' (Domínguez-García et al. (2011)). We have also developed Brownian dynamics simulations to be compared with the experiments. The Fig.4 summarizes some of our results in aggregation and disaggregation. The experimental kinetic exponents during disaggregation range from $z = 0.44$ to 1.12 and $z' = 0.27$ to 0.67, while simulations give very regular values, with z and $z' \sim 1$. Then, the kinetic exponents do not agree, being also the process of disaggregation much faster in simulations. From these results, we conclude that remarkable differences exist between a simple theoretical model and the interactions in our experimental setup, differences that are specially important when the influence of the applied magnetic field is removed. In all these experiments some data has been collected before any external field is applied. That allows us to study the microstructure of the suspensions by calculating the electrostatics potential using the methods previously explained. The inversion of the O-Z equation reveals an attractive well in the potential with a value in its minimum in the order of $-0.2\,k_BT$, similar to other observations of attractive interactions of sedimented particles in confinement situations. Moreover, these values of the minimum in the potential seems to depend of the concentration of particles (Domínguez-García, Pastor, Melle & Rubio (2009)), something which is expected, if it is related in some way with the electrical charge contained in the suspension.

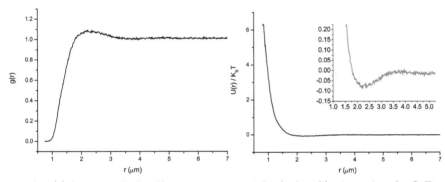

Fig. 5. Left: $g(r)$ function, Right: Electrostatic potential calculated by inverting the O-Z equation (all the approximations give the same result) Inset: a detail for $U(r)$ in the region of the minimum. Number density $n = 0.0009$

As a confirmation of these results, we show here a calculation of the electrostatic potential using a long set of images of charged superparamagnetic microparticles spreading in the experimental system described above. We have obtained images of the suspension during more than an hour, with a temporal lapse between images of 0.3 seconds. This data allow us to produce a very defined graph for the pair correlation function, showed in the Fig.5. In the right side of the Fig.5, we plot the electrostatic potential and in its inset we can see that the minimum has a value of about $-0.1\,k_BT$, confirming the previous results obtained in this experimental setup.

However, this result may be an effect of an imagining artefact. About that question, some of the studies which use particle tracking only apply some filters to the images for detecting brightness points and then extracting the position of the particles. Our image analysis software (Domínguez-García & Rubio (2009)) employs open-sourced algorithms for detecting the centres of mass of the particles by detecting the borders of each object and then obtaining its geometrical properties. As an example, we have tried to evaluate how this border detection can have an influence on the result of the electrostatic potential. A measured apparent displacement $\Delta(r) = r' - r$ should affect to the radial distribution function in the following form: $g(r) = g'(r + \Delta(r))(1 + d\Delta(r)/dr)$ (Polin et al. (2007)). From that expression, the variation in the electrostatic potential is:

$$\beta U'(r) - \beta U(r) \cong -\beta \frac{dU(r)}{dr}\Delta(r) + \frac{d\Delta(r)}{dr} \tag{48}$$

For obtaining $\Delta(r)$ we have extracted a typical particle image and we have composed some set of images which consist on separating the two particles a known distance (r) in pixels. Next, we apply our methods of image analysis for obtaining the position of those particles and calculate the distances (r'). Then, the apparent displacement, $\Delta(r) = r' - r$, is observed to grow when the particles are very near. In Fig.6, we display the results of our calculations on the possible artefact in the analysis of the position of the particles by image binarization and binary watershed, a method for automatically separating particles that are in contact. The figure reveals that the correction on the electrostatic potential for this cause is basically negligible, because the correction in the potential is zero for distances $r > 1.2\ \mu$m. In the inset of the figure we can see some of the images we have employed for this calculation, showing the detected border of the particles among the images themselves.

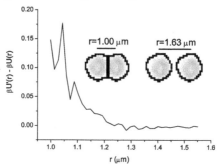

Fig. 6. Estimation of a possible artefact in the analysis of the position of the particles. In the inset we have included some examples of the images used for this calculation.

In any case, the possibility of an artifact can be the cause of these observations in the electrostatic potential cannot be descarted. However, the direct or indirect presence and influence of these attractive wells has been detected in many other situations in these experiments. For example, the attractive interaction disappears when we added a salt, in our case KCl, to the suspensions, confirming the electrostatic nature of the phenomena (Domínguez-García, Pastor, Melle & Rubio (2009)). In disaggregation it is observed how the particles move inside the chains without leaving them (Domínguez-García et al. (2011)). The lapse of time that the particles are in this situation depends on the initial morphology of the aggregates, something which has been observed to depend on the ratio R_1/R_0 (Domínguez-García, Melle & Rubio (2009); Domínguez-García & Rubio (2010)). Then, this effective lapse of time depends of how many particles are located near the other in a short distance. In that situation, the attractive interaction should play a role in disaggregation, as it eems to be. Indeed, this "detaining" effect of the particles inside the clusters is not observed n experiments with added salt.

Vhat is more, we have also observed that the kinetic exponents during aggregation are lifferent and slower if we add salt to the suspension (Domínguez-García et al. (2011)). This ast effect may be related with an unexpected interaction of the particles with the charged uartz bottom wall by means of a a spontaneous macroscopic electric field. When the particles nd clusters have no electrical component, they should be highly sedimented at the bottom of he quartz cell and the resistance to their the movement should be increased (Kutthe (2003)), enerating that the kinetic exponents reduce their value.

4. Conclusions

n this chapter, we have reviewed the main interactions, with focus on hydrodynamics and rom a experimental point of view, that can be important in a confined colloidal system at low oncentration of microparticles. We have used charged superparamagnetic microparticles lispersed in water in low-confinement conditions by means of a glass cell for the study of irreversible field-induced aggregation and disaggregation, as well as the microstructure of the suspension. Regarding aggregation characteristic times and basic behaviour on the lisaggregation of the particles, we have observed significant discrepancies between the xperimental results and the theory. Morover, anomalous effects in the electrostatic behaviour lave been observed, showing that, in this kind of systems, the electro-hydrodynamics nteractions are not well understood at present and deserve more theoretical and experimental nvestigations.

5. Acknowledgements

Ve wish to acknowledge Sonia Melle and J.M. Pastor for all the work done, J.M M. González, J.M. Palomares and F. Pigazo (ICMM) for the VSM magnetometry measurements nd J.C. Gómez-Sáez for her proofreading of the English texts. This research has been artially supported by M.E.C. under Project No. FIS2006-12281-C02-02, M.C.I under IS2009-14008-C02-02, C.A.M under S/0505/MAT/0227 and UNED by 2010V/PUNED/0010.

6. References

Andreu, J. S., Camacho, J. & Faraudo, J. (2011). Aggregation of superparamagnetic colloids in magnetic fields: the quest for the equilibrium state, 7: 2336.

Bacri, J. C., Djerfi, K. & Neveu, S. (1993). Ferrofluid viscometer - transient magnetic birrefrigence in crossed fields, *J. Mag. Mat.* 123 (1-2): 67–73.

Baumgart, J., Arauz-Lara, J. L. & Bechinger, C. (2006). Like-charge attraction in confinement: myth or truth?, *Soft Matter* 2: 631–635.

Behrens, S. H. & Grier, D. G. (2001a). The charge of glass and silica surfaces, *J. Chem. Phys.* 115: 6716–6721.

Behrens, S. H. & Grier, D. G. (2001b). Pair interaction of charged colloidal spheres near a charged wall, *Phys. Rev. E* 64: 050401(R).

Bensech, T. & Yiacoumi, S. (2003). Brownian motion in confinement., *Phys. Rev. E* 68: 021401.

Bossis, G., Mathis, C., Minouni, Z. & Paparoditis, C. (1990). Magnetoviscosity of micronic particles, *Europhys. Lett.* 11: 133.

Brenner, H. (1961). The slow motion of a sphere through a viscous fluid towards a plane surface., *Phys. Rev. E* 68: 021401.

Buzzaccaro, S., Tripodi, A., Rusconi, R., Vigolo, D. & Piazza, R. (2008).   Kinetics of sedimentation in colloidal suspensions, *J. Phys. Condens. Matter* (20): 494219.

Carrique, F. & Ruiz-Reina, E. (2009). Effects of water dissociation and co2 contamination on the electrophoretic mobility of a spherical particles in aqueous salt-free concentrated suspensions, *J. Phys. Chem. B* (113): 8613–8625.

Cernak, J. (1994). Aggregation of needle-like macro-clusters in thin-layers of magnetic fluid *J. Magn. Magn. Mater.* 132: 258.

Cernak, J., Helgesen, G. & Skjeltorp, A. T. (2004). Aggregation dynamics of nonmagnetic particles in a ferrofluid, *Phys. Rev. E* 70 (3): 031504 Part 1.

Cernak, J., Macko, P. & Kasparkova, M. (1991).   Aggregation and growth-processes in thin-films of magnetic fluid, *J. Phys. D: Appl. Phys.* 24: 1609.

Chen, H. & Ma, H. (2006). The density profile of hard sphere liquid system under gravity., *J Chem. Phys.* 125: 024510.

Crocker, J. C. (1997). Measurement of the hydrodynamic corrections of the Brownian motion of two colloidal spheres., *J. Chem. Phys.* 106: 2837–2840.

Crocker, J. C. & Grier, D. G. (1996). Methods of digital video microscopy for colloidal studies *J. Colloid Interface Sci.* 179: 298–310.

Crocker, J. C., Valentine, M. T., Weeks, E. R., Gisler, T., Kaplan, P. D., Yodh, A. G. & Weitz, D. A. (2000). Two-point microrheology of inhomogeneous soft materials, *Phys. Rev. Lett.* 85: 888–891.

Derjaguin, B. V. & Landau, L. (1941). Theory of the stability of strongly charged lyophobic sols and of the adhesion of strongly charged particles in solution of electrolytes., *Acta Physicochim. URSS* 14: 633.

Doi, M. & Edwards, S. (1986). *The Theory of Polymer Dynamics.*, Clarendon Press, Oxford.

Domínguez-García, P., Melle, S., Pastor, J. M. & Rubio, M. A. (2007). Scaling in the aggregation dynamics of a magneto-rheological fluid., *Phys. Rev. E* 76: 051403.

Domínguez-García, P., Melle, S. & Rubio, M. A. (2009). Morphology of anisotropic chains in a magneto-rheological fluid during aggregation and disaggregation processes., *J Colloid Interface Sci.* 333: 221–229.

Domínguez-García, P., Pastor, J. M., Melle, S. & Rubio, M. A. (2009). Electrostatic and hydrodynamics effects in a sedimented magneto-rheological suspension, *Phys. Rev. E* 80: 0214095.

Domínguez-García, P., Pastor, J. M. & Rubio, M. A. (2011). Aggregation and disaggregation dynamics of sedimented and charged superparamagnetic microparticles in water suspension, *Europhys. J. E. Soft. Matter.* 34: 36.

Domínguez-García, P. & Rubio, M. A. (2009). Jchainsanalyser: an imagej-based stand-alone application for the study of magneto-rheological fluids., *Comput. Phys. Commun.* 80: 1956–1960.

Domínguez-García, P. & Rubio, M. A. (2010).   Three-dimensional morphology of field-induced chain-like aggregates of superparamagnetic micro-particles, *Colloids Surf., A* 358: 21–27.

Dufresne, E. R., Altman, D. & Grier, D. G. (2001). Brownian dynamics of a sphere between parallel walls, *Europhys. Lett.* 53: 264–270.

Dufresne, E. R., Squires, T. M., Brenner, M. P. & Grier, D. G. (2000). Hydrodynamic coupling of two brownian spheres to a planar surface, *Phys. Rev. Lett.* 85(15): 3317–3320.

Egatz-Gómez, A., Melle, S., García, A. A., Lindsay, S. A., Márquez, M., Domínguez-García, P., Rubio, M. A., Picraux, S. T., Taraci, J. L., Clement, T., Yang, D., Hayes, M. A. & Gust, D. (2006). Discrete magnetic microfluidics, *Appl. Phys. Lett.* 89(3)(034106).

Faucheux, L. P. & Libchaber, A. J. (1994). Confined brownian motion, *Phys. Rev. E* 49: 5158.

Fermigier, M. & Gast, A. P. (1992). Structure evolution in a paramagnetic latex suspension, *J. Colloid Interface Sci.* 154: 522.

Fraden, S., Hurd, A. J. & Meyer, R. B. (1989). Electric-field-induced association of colloidal particles, *Phys. Rev. Lett.* 63 (21): 2373–2376.

Furst, E. M. (2005). Applications of laser tweezers in complex fluid rheology., *Curr. Opin. Colloid Interface Sci.* 10: 79–86.

Furst, E. M. & Gast, A. P. (2000). Micromechanics of magnetorheological suspensions, *Phys. Rev. E* 61: 6732.

González, A. E., Odriozola, G. & Leone, R. (2004). Colloidal aggregation with sedimentation: concentration effects., *Europhys. J. E. Soft. Matter.* 13: 165–178.

Grier, D. G. (2003). A revolution in optical manipulation., *Nature* 424: 810–815.

Grier, D. G. & Han, Y. (2004). Anomalous interactions in confined charge-stabilized colloid, *J. Phys. Condens. Matter* 16: 4145–4157.

Gyger, M., Rückerl, F., Käs, J. A. & Ruiz-García, J. (2008). Errors in two particle tracking at close distances, *J. Colloid Interface Sci.* 326: 382–386.

Halsey, T. C. & Toor, W. (1990). Structure of electrorheological fluids, *Phys. Rev. Lett.* 65: 2820.

Han, Y. & Grier, D. G. (2003). Confinement-induced colloidal attractions in equilibrium, *Phys. Rev. Lett.* 91(3): 038302.

Helgesen, G., Pieranski, P. & Skjeltorp, A. T. (1990). Nonlinear phenomena in systems of magnetic holes, *Phys. Rev. Lett.* 64 (12): 1425–1428.

Helgesen, G., Skjeltorp, A. T., Mors, P. M., Botet, R. & Jullien, R. (1988). Aggregation of magnetic microspheres: Experiments and simulations, *Phys. Rev. Lett.* 61(15): 1736–1739.

Hernández-Ortiz, J. P., Ma, H., de Pablo, J. J. & Graham, M. D. (2006). Cross-stream-line migration in confined flowing polymer solutions: Theory and simulation., *Phys. Fluids* 18: 123101.

Jendrejack, R. M., Schwartz, D. C., Graham, M. D. & de Pablo, J. J. (2003). Effect of confinmeent on DNA dynamics in microfluidic devices., *J. Chem. Phys.* 119(2): 1165–1173.

Keller, M., Schilling, J. & Sackmann, E. (2001). Oscillatory magnetic bead rheometer for complex fluid microrheometry, *Rev. Sci. Instr.* 72(9): 3626.

Kerr, R. A. (1990). *Science* 247: 050401.

Kolb, M., Botet, R. & Jullien, R. (1983). Scaling of kinetically growing clusters, *Phys. Rev. Lett.* 51: 1121–1126.

Komeili, A. (2007). Molecular mechanisms of magnetosome formation., *Annu. Rev. Biochem.* 76: 351–356.

Koppel, D. E. (1972). Analysis of macromolecular polydispersity in intensity correlation spectroscopy: The method of cumulants., *J. Chem. Phys.* 57(11): 4814–4820.

Kutthe, R. (2003). Stokesian dynamics of nonspherical particles, chains, and aggregates., *J. Chem. Phys.* 119(17): 9280–9294.

Larsen, A. E. & Grier, D. G. (1997). Like-charge attraction in metastable colloidal crystallites, *Nature* 385(16): 230–233.

Larson, R. G. (1999). *The Structure and Rheology of Complex Fluids*, Oxford University Press, New York.

Lian, Z. & Ma, H. (2008). Electrostatic interaction between two nonuniformly charged colloid particles confined in a long charged cylinder wall., *J. Phys. Condens. Matter* 20: 035109.

Lin, B., Yu, J. & Rice, S. A. (2000). Direct measurements of constrained Brownian motion of an isolated sphere between two walls., *Phys. Rev. E* 62 (3): 3909–3919.

Lin, M. Y., Lindsay, H. M., Weitz, D. A., Ball, R. C., Klein, R. & Meakin, P. (1989). Universality in colloid aggregation, *Nature* 339(1): 360–362.

Liron, N. & Mochon, S. (1976). Stokes flow for a stokeslet between two parallel flat plates., *J. Eng. Math.* 10 (4): 287–303.

Lord Corporation, http://www.lord.com/ (n.d.).

Martin, J. E. & Anderson, R. A. (1996). Chain model of electrorheology, *J. Chem. Phys.* 104: 4814.

Martin, J. E., Odinek, J. & Halsey, T. C. (1992). Evolution of structure in a quiescent electrorheological fluid, *Phys. Rev. Lett.* 69 (10): 1524–1527.

Martínez-Pedrero, F., Tirado-Miranda, M., Schmitt, A. & Callejas-Fernández, J. (2007). Formation of magnetic filaments: A kinetic study., *Phys. Rev. E* 76: 011405.

Mason, T. G. (2000). Estimating the viscoelastic moduli of complex fluid using the generalized stokes-einstein equation, *Rheol. Acta* 39: 371–378.

Mason, T. G. & Weitz, D. A. (1995). Optical measurements of frequency-dependent linear viscoelastic moduli of complex fluids, *Phys. Rev. Lett.* 74: 1250 – 1253.

Meakin, P. (1983). Formation of fractal clusters and networks by irreversible diffusion-limited aggregation, *Phys. Rev. Lett.* 51: 1119–1122.

Melle, S. (2002). *Estudio de la dinámica de suspensiones magneto-reológicas sometidas a campos externos mediante el uso de técnicas ópticas. Procesos de agregación, formación de estructuras y su evolución espacio-temporal.*, PhD thesis, Universidad Nacional de Educación a Distancia.

Melle, S., Rubio, M. A. & Fuller, G. G. (2001). Time scaling regimes in aggregation of magnetic dipolar particles: scattering dichroism results., *Phys. Rev. Lett.* 87(11): 115501.

Miguel, M. C. & Pastor-Satorras, R. (1999). Kinetic growth of field-oriented chains in dipolar colloidal solutions, *Phys. Rev. E* 59 (1): 826–834.

Miyazima, S., Meakin, P. & Family, F. (1987). Aggregation of oriented anisotropic particles, *Phys. Rev. A* 36 (3): 1421–1427.

Nakano, M. & Koyama, K. (eds) (1998). *Proceedings of the 6th International Conferences on ER and MR fluids and their applications*, World Scientific, Singapore.

Odriozola, G., Jiménez-Ángeles, F. & Lozada-Cassou, M. (2006). Effect of confinement on the interaction between two like-charged rods., *Phys. Rev. Lett.* 97: 018102.

Otterstedt, J. & Brandreth., D. A. (1998). *Small Particles Technology.*, Springer.

Paddinga, J. T. & Briels, W. J. (2010). Translational and rotational friction on a colloidal rod near a wall, *J. Chem. Phys.* 132: 054511.

Pailthorpe, B. A. & Russel, W. B. (1982). The retarded van der Waals interaction between spheres, *J. Colloid Interface Sci.* 89(2): 563–566.

Philipse, A. P. & Koenderink, G. H. (2003). Sedimentation-diffusion profiles and layered sedimentation of charged colloids at low ionic strength, *Adv. Colloid Interface Sci.* 100–102: 613–639.

Polin, M., Grier, D. G. & Han, Y. (2007). Colloidal electrostatic interactions near a conducting surface., *Phys. Rev. E* 76: 041406.

Promislow, J., Gast, A. P. & Fermigier, M. (1994). Aggregation kinetics of paramagnetic colloidal particles, *J. Chem. Phys.* 102(13): 5492–5498.

Rabinow, J. (1948). The magnetic fluid clutch, *AIEE Trans.* 67: 1308.

ydrodynamics on Charged Superparamagnetic Microparticles in Water Suspension: Effects of Low-Confinement
onditions and Electrostatics Interactions

79

.asa, M. & Philipse, A. P. (2004). Evidence for a macroscopic electric field in the sedimentation profiles of charged colloids, *Nature* 429(24): 857–860.

.einer, E. S. & Radke, C. J. (1993). Double layer interactions between charge-regulated colloidal surfaces: Pair potentials for spherical particles bearing ionogenic surfaces groups., *Adv. Colloid Interface Sci.* 47: 59–147.

.ussel, W. B., Saville, D. A. & Schowalter, W. R. (1989). *Colloidal Dispersions.*, Cambridge University Press.

.avin, T. & Doyle, P. S. (2005). Static and dynamic error in particle tracking microrheology, *Biophysical Journal* 88: 623–638.

.avin, T. & Doyle, P. S. (2007). Statistical and sampling issues when using multiple particle tracking, *Phys. Rev. E* 76: 021501.

.avin, T., Spicer, P. T. & Doyle, P. S. (2007). A rational approach to noise characterization in video microscopy particle tracking, *Phys. Rev. E* 76: 021501.

.chmidt, M., Dijkstra, M. & Hansen, J. P. (2004). Competition between sedimentation and phase coexistence of colloidal dispersions under gravity, *J. Phys. Condens. Matter* 16: S4185–S4194.

.chmitz, K. S., Bhuiyan, L. B. & Mukherjee, A. K. (2003). On the grier-crocker/tata-ise controversy on the macroion-macroion pair potential in a salt-free colloidal suspension, *Langmuir* 19: 7160–7163.

.cience. (1999). Complex systems., *Science* 284(5411): 1–212.

.hen, L., Stachowiak, A., Fateen, S. E. K., Laibinis, P. E. & Hatton, T. A. (2001). Structure of alkanoic acid stabilized magnetic fluids. A small-angle neutron and light scattering analysis, *Langmuir* 17: 288.

.kjeltorp, A. T. (1983). One-dimensional and two-dimensional crystallization of magnetic holes, *Phys. Rev. Lett.* 51 (25): 2306–2309.

.mirnov, P., Gazeau, F., Lewin, M., Bacri, J. C., Siauve, N., Vayssettes, C., Cuenod, C. A. & Clement, O. (2004). In vivo cellular imaging of magnetically labeled hybridomas in the spleen with a 1.5-t clinical mri system, *Magn. Reson. Medi.* 52: 73–79.

.quires, T. M. & Brenner, M. P. (2000). Like-charge attraction and hydrodynamic interaction, *Phys. Rev. Lett.* 85(23): 4976–4979.

.ao, R. (ed.) (2000). *Proceedings of the 7th International Conferences on ER and MR fluids*, World Scientific, Singapore.

.ata, B. V. R. & Ise, N. (1998). Monte carlo study of structural ordering in charged colloids using a long-range attractive interaction, *Phys. Rev. E* 58(2): 2237–2246.

.ata, B. V. R. & Ise, N. (2000). Reply to "comment on 'monte carlo study of structural ordering in charged colloids using a long-range attractive interaction' ", *Phys. Rev. E* 61(1): 983–985.

.ata, B. V. R., Mohanty, P. S. & Valsakumar, M. C. (2008). Bound pairs: Direct evidence for long-range attraction between like-charged colloids, *Solid State Communications* 147: 360–365.

.irado, M. M. & García, J. (1979). Translational friction coeffcients of rigid, symmetric top macromolecules. Application to circular cylinders., *J. Chem. Phys.* 71: 2581.

.irado, M. M. & García, J. (1980). Rotational dynamics of rigid symmetric top macromolecules. Application to circular cylinders., *J. Chem. Phys.* 73: 1986.

.irado-Miranda, M. (2001). *Agregación de Sistemas Coloidales Modificados Superficialmente.*, PhD thesis, Universidad de Granada.

U. S. National Institutes of Health, Bethesda, Maryland, USA, http://rsb.info.nih.gov/ij/ (n.d.).

Verdier, C. (2003). Rheological properties of living materials., *J. Theor. Medic.* 5: 67–91.

Verwey, E. J. W. & Overbeek, J. T. G. (1948). *Theory of the Stability of Lyophobic Colloids.*, Elsevier Amsterdam.

Vicsek, T. (1992). *Fractal Growth Phenomena*, 2 edn, World Scientific, Singapore.

Vicsek, T. & Family, F. (1984). Dynamic scaling for aggregation of clusters, *Phys. Rev. Lett.* 52,19: 1669–1672.

von Smoluchowski, M. (1917). *Z. Phys. Chem., Stoechiom. Vertwanddtschaftsl* 92: 129.

Vuppu, A. K., García, A. A., Hayes, M. A., Booksh, K., Phelan, P. E., Calhoun, R. & Saha, S. K. (2004). Phase sensitive enhancement for biochemical detection using rotating paramagnetic particle chains, *J. Appl. Phys.* 96: 6831–6838.

Waigh, T. A. (2005). Microrheology of complex fluids, *Rep. Prog. Phys.* 68: 685–742.

Wilhelm, C., Browaeys, J., Ponton, A. & Bacri, J. C. (2003). Rotational magnetic particles microrheology: The maxwellian case, *Phys. Rev. E* 67: 011504.

Wilhelm, C., Gazeau, F. & Bacri, J. C. (2003). Rotational magnetic endosome microrheology: Viscoelastic architecture inside living cells, *Phys. Rev. E* 67: 061908.

Wilhelm, C., Gazeau, F. & Bacri, J. C. (2005). Magnetic micromanipulation in the living cell, *Europhys. news* 3: 89.

Witten, T. A. & Sander, L. M. (1981). Diffusion-limited aggregation, a kinetic critical phenomenon, *Phys. Rev. Lett.* 47: 1400–1403.

4

Magnetohydrodynamics of Metallic Foil Electrical Explosion and Magnetically Driven Quasi-Isentropic Compression

Guiji Wang, Jianheng Zhao, Binqiang Luo and Jihao Jiang
Institute of Fluid Physics, China Academy of Engineering Physics,
Mianyang City, Sichuan Province
China

1. Introduction

The electrical explosion of conductors, such as metallic foils and wires, refers to rapid changes of physical states when the large pulsed current (tens or hundreds of kA or more, the current density $j \geq 10^6$ A/cm²) flows through the conductors in very short time(sub microsecond or several microseconds), which may produce and radiate shock waves, electrical magnetic waves, heat and so on. There are many applications using some characteristics of the electrical explosion of conductors.

The Techniques of metallic foil electrical explosion had been developed since 1961, which was first put forward by Keller, Penning[1] and Guenther et al[2]. However, it develops continually until now because of its wide uses in material science, such as preparation of nanometer materials and plating of materials[3,4], shock wave physics[5-7] , high energy density physics[8] and so on. Especially the techniques of metallic foil electrically exploding driving highvelocity flyers, are widely used to research the dynamics of materials, hypervelocity impact phenomena and initiation of explosives in weapon safety and reliability. Therefore, in this chapter we focus on the physical process of metallic foil explosion and the techniques of metallic foil electrically exploding driving highvelocity flyers. Here the explosion of metallic foils are caused by the large current flowing through in sub microsecond or 1~2 microsecond or less. During the whole physical process, not only does the temperature rising, melting, vaporizing and plasma forming caused by instantaneously large current, but also the electrical magnetic force exists and acts on. Because the whole process is confined by rigid face and barrel, and the time is very short of microsecond or sub microsecond or less, and the phynomena is similar to the explosion of explosives, we call the process electrical explosion of metallic foils. This process is a typically hydrodynamic phenomena. It is also a magnetohydrodynamic process because of the exist and action of the magnetic force caused by large current and self-induction magnetic field.

Magnetically driven quasi-isentropic compression is an relatively new topic, which was developed in 1972[9]. At that time the technique of magnetically driven quasi-isentropic compression was used to produce high pressure and compress the cylindrical sample materials. Until 2000, the planar loading technique of magnetically driven quasi-isentropic compression was firstly presented by J.R. Asay at Sandia National Laboratory[10]. In las decade, this planar loading technique had being developed fastly and accepted by man* researchers in the world, such as France[11], United Kingdom[12],and China[13]. As J.R. Asa* said, it will be a new experimental technique widely used in shock dynamics, astrophysics high energy density physics, material science and so on. The process of magnetically drivel quasi-isentropic compression is typical magnetodynamics[14], which refers to dynami* compression, magnetic field diffusion, heat conduction and so on.

As described above, the electrical explosion of metallic foil and magnetically driven quasi isentropic compression is typically magnetohydrodynamic problem. Although it develop* fastly and maybe many difficulties and problems exist in our work, we present ou* important and summary understanding and results to everyone in experiments anc simulations of electrical explosion of metallic foil and magnetically driven quasi-isentropi* compression in last decade.

In the following discussions, more attentions are paid to the physical process, th* experimental techniques and simulation of electrical explosion of metallic foil anc magnetically driven quasi-isentropic compression.

2. Physical process of metallic foil electrical explosion and magnetically driven quasi-isentropic compression

2.1 Metallic foil electrical explosion

Here we introduce the model of metallic foil electrically exploding driving highvelocity flyers to describe the physical process of electrical explosion of metallic foil shown in Fig.1 A large pulsed current is released to the metallic foil of the circuit, which is produced by a typically pulsed power generator. The circuit can be described by R-C-L electrical circui* equations[15]. During the circuit, the metallic foil is with larger resistance than that of othe* part, so the energy is mainly absorbed by the metallic foil, and then the physical states o* metallic foil change with time. Fig.2 shows the typical current and voltage histories betweer metallic aluminum foil during the discharging process of pulsed power generator.

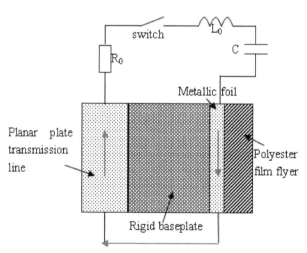

Fig. 1. The model of metallic foil electrically exploding driving highvelocity flyers.

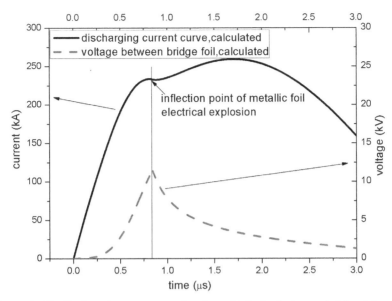

Fig. 2. The typically discharging current and voltage histories between bridge Aluminum foil.

According to the density changing extent of metallic foil when the first pulsed current flows through it, the whole process of electrical explosion of metallic foil can be classified to two stages. The initial stage includes the heating stage , the melting stage and the heating stage of liquid metal before vaporizing. During this process, the density of metallic foil changes relatively slow. The second stage includes the vaporizing stage and the following plasma forming. The typical feature of electrical explosion of metallic foil is that the foil expands rapidly and violently, and that the resistance increases to be two or more orders than that of initial time $(R/R_0 \sim 100)$. The resistance increases to be maximum when the state of metallic foil is at the vaporizing stage. During this stage, the voltage of between foil also increases to be maximum, and then the breakdown occurs and the plamas is forming. The inflection point of the discharging current shown in Fig.2 exhibits the feature.

At the initial satge, the expansion of metallic foil is not obvious, and the change of physical states can be described with one thermodynamic variable T (temperature) or specific enthalpy. The energy loss of the interaction between the foil and the ambient medium can be neglected when there is no surface voltaic arcs. Therefore, some assumptions can be used to simplify the problem. We can think that the heating of the metallic foil is uniform and the instability, heat conduction and skin effect can not be considered at initial stage. For this stage, the physical states of metallic foil vary from solid to liquid, and the model of melting phase transition can be used to described it well[16].

For the second stage, the physical states varies from liquid to gas, and then from gas to plasma. There are several vaporizing mechanisms to describe this transition, such as surface evaporation and whole boil[16]. The rapid vaporizing of liquid metal make its resistance increases violently, and the current decreases correspondingly. At this time, the induction voltage between bridge foil increases fastly. If the induction voltage can make the metallic vapor breakdown and the plasma is formed, the circuit is conducted again. Of course, the breakdown of metallic vapor needs some time, which is called relaxation time as shown in Fig.3. For different charging voltages, the relaxation time varies, which can be seen from the experimental current hostories in Fig.3.

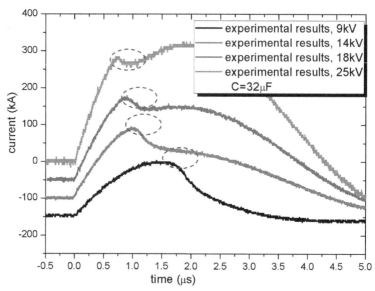

Fig. 3. The breakdown relaxation time shown in the discharging current histories at different charging voltage for the pulsed power generator.

One important application of the electrical explosion of metallic foil is to launch highvelocity flyers with the rapid expansion of tha gas and plasma from electrical explosion of metallic foil. Some metallic materials are with good conductivity and explosion property, such as gold, silver, copper, aluminum and so on. The experimental results[17] show that the aluminum foil is the best material for the application of metallic foil electrically exploding driven highvelocity flyers. There are many models used to describe the process, such as eletrical Gurney model[18], Schmidt model[19] and one dimensional magnetohydrodynamic model[20]. The electrical Gurney model and Schmidt model are two empirical models which are derived from energy conservation equation based on some assumptions. For a specific electrical parameters of the circuit of some apparatus, the electrical Gurney model can be used to predict the final velocity of the flyers when the Gurney parameters are determined based on some experimental results. And the Schmidt model can be used to predict the velocity history of the flyers because the Gurney energy part is substituted with an energy part with the function of time, which is depended on the measured current and voltage histories between bridge foil to correct the specific power coefficient. These two models can't reflect other physical variables of electrical explosion of metallic foil except the velocity of the flyer. Therefore, a more complex model is put forward based on magnetohydrodynamics, which considers heat conduction, magnetic pressure and electrical power. The magnetohydrodynamic model can well reflect the physical process of electrical explosion of metallic foil. The equations are given below[16,20].

$$\begin{cases} \dot{x} = u; \quad \dot{v} = \dfrac{\partial}{\partial q}(x^{\gamma-1}u) \\[2mm] \dot{u} + x^{\gamma-1}\dfrac{\partial}{\partial q}\left(p + p_\omega + \dfrac{B^2}{2\mu_0}\right) = 0 \\[2mm] \dot{\epsilon} + (p + p_\omega)\dot{v} = Q_v \\[2mm] \dfrac{d}{dt}(x^{1-\gamma}\,v\,B) = \dfrac{\partial E}{\partial q} \\[2mm] E = \dfrac{1}{\mu_0 \sigma\, v}\dfrac{\partial}{\partial q}(x^{\gamma-1}B) \\[2mm] j = \sigma E; \quad Q_v = v\,jE \\[2mm] p = p(v,T); \quad \epsilon = \epsilon(v,T); \quad \sigma = \sigma(v,T) \end{cases} \tag{1}$$

Where, γ—symmetric exponent (for metallic wire or cylindrical foil $\gamma = 2$，and for planar foil $\gamma = 1$); $\partial/\partial q = x^{1-\gamma}v\partial/\partial x$; q—Lagrange mass coordinate;B—transverse component of magnetic field;E—axial component of electrical field; j—current density; Q_v—specific power of Joule heating; p_ω—artificial viscosity coefficient；u—transverse moving velocity；p—pressure;ϵ—internal energy; v—unit volume; σ—conductivity. For this apparatus, the discharging ciruit is a typical RCL circuit, which can be expressed by equation (2) below.

$$\begin{cases} \dfrac{d}{dt}\big[(L_0 + L_{foil})I\big] + U_{foil} + R_0 I = U_c; \\[2mm] \dfrac{dU_C}{dt} = -\dfrac{I}{C_0}; \\[2mm] U_{foil} \simeq l_{foil}E[t,X(t)]\ , \end{cases} \tag{2}$$

In the equation (2), when the time $t = 0$, the primary current and voltage $I(0) = 0$ and $U_c(0) = U_0$, C_0 and U_0 are the capacitance and charging voltage of capacitor or capacitor bank, L_0 and R_0 are the inductance and efficient resistance of circuit, U_{foil} is the voltage between the ends of metallic foil, which is related with the length l_{foil} of metallic foil and the magetic field of the space around the foil. the dynamic inductance L_{foil} can be obtained by equation (3).

$$L_{foil}(t) = \mu_0 k(l_{foil}/b)\left[x_0' - X_0\right] \tag{3}$$

Where μ_0 is the vacuum magnetic permeability, k is a coefficient related with the length l and width b of metallic foil. x is the expanding displacement of metallic foil.

2.2 Magnetically driven quasi-isentropic compression

The concept of magnetically driven quasi-isentropic compression is illustrated in Fig.4. A direct short between the anode and cathode produces a planar magnetic field between the conductors when a pulsed current flows through the electrodes over a time scale of 300～800ns. The interaction between the current (density J) and the induction magnetic field

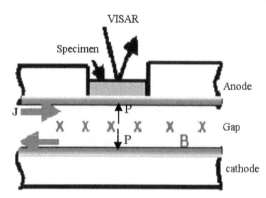

Fig. 4. The principle diagram of magnetically driven quasi-isentropic compression.

B produces the magnetic pressure ($\vec{J} \times \vec{B}$) proportional to the square of the field. The force is loaded to the internal surface that the current flows through. The loading pressure wave is a ramp wave, which is a continuous wave. Compared with the shock wave, the increment of temperature and entropy is very lower. However, because of the effects of viscosity and plastic work, the sample can't turn back to the original state after the loading wave. That is to say, in solids the longitudinal stress differs from the hydrostatic pressure because of resolved shear stresses that produce an entropy increase from the irreversible work done by deviator[21, 22]. For this reason, the ramp wave loading process is usually assumed to be quasi-isentropic compression. Besides the loading force is magnetic pressure, it is called magnetically driven quasi-isentropic compression.

In order to produce high pressure, the amplitude of the current is ususally up to several megamperes or tens of megamperes. Because of the effects of Joule heating and magnetic field diffusion, the physical states of the loading surface will change from solid to liquid, and to gas and plasma. And these changes will propagate along the thickness direction of the electrodes originated from the loading surface. These phenomena are typically magnetohydrodynamic problems. In order to describe the physical process, the equation of magnetic field diffusion is considered besides the equations of mass, momentum and energy. The magnetohydrodynamic equations are presented below.

$$
\left\{
\begin{aligned}
& \frac{\partial \rho_m}{\partial t} + \nabla \cdot (\rho_m \vec{u}) = 0 \\
& \rho_m \frac{d\vec{u}}{dt} + \nabla(p+q) - \vec{J} \times \vec{B} = 0 \\
& \frac{de}{dt} + (p+q)\frac{d(1/\rho_m)}{dt} - \dot{e}_D = 0 \\
& \rho_m \frac{d}{dt}\left(\frac{\vec{B}}{\rho_m}\right) - (\vec{B} \cdot \nabla)\vec{u} = -\nabla \times \left[\frac{\eta}{\mu_0}(\nabla \times \vec{B})\right] \\
& \vec{J} = \sigma\vec{E} = \frac{1}{\mu_0}\nabla \times \vec{B} \\
& \vec{u} = \frac{d\vec{x}}{dt}, \dot{e}_D = \kappa\nabla T
\end{aligned}
\right.
\tag{4}
$$

Where ρ_m is mass density of electrodes, u is velocity, J is current density, B is magnetic field, is pressure, q is artificial viscosity pressure, e is specific internal energy, σ is electrical conductivity of electrodes and κ is thermal conducitivity. imilar to the technique of electrical explosion of metallic foil, the large current is also roduced by some pulsed power generators, for example, the ZR facility at Sandia National aboratory can produce a pulsed current with peak value from 16 MA to 26 MA and rising me from 600 ns to 100 ns[23]. In the following part, we will introduce the techniques of magnetically driven quasi-isentropic compression based on the pulsed power generators eveloped by ourselves.

. Techniques of metallic foil electrically exploding driving highvelocity flyers nd magnetically driven quasi-isentropic compression

he techniques of metallic foil electrically exploding driving highvelocity flyers and magnetically driven quasi-isentropic compression have been widely used to research the ynamic properties of materials and highvelocity impact phenomena in the conditions of hock and shockless(quasi-isentropic or ramp wave) loading. By means of these two echniques, we can know the physical, mechnical and thermodynamic properties of naterials over different state area (phase space), such as Hugoniot and off-Hugoniot states.

.1 Metallic foil electrically exploding driving highvelocity flyers[24,25,26]

As descibed above, the high pressure gas and plasma are used to launch highvelovity flyer plates, which are produced from the electrical explosion of metallic foil. The working principle diagram of the metallic foil electrically exploding driving highvelocity flyers is resented in Fig.5. Usually we choose the pure aluminum foil as the explosion material because of its good electrical conductivity and explosion property. The flyers may be olyester films, such as Mylar or Kapton, or complex ones consisted of polyester film and netallic foil. The material of barrel for accelerating the flyers may be metals or non-polyester ilms, such as Mylar or Kapton, or complex ones consisted of polyester film and metallic foil. he material of barrel for accelerating the flyers may be metals or non-metals, such as

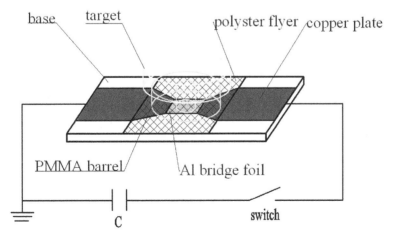

ig. 5. The diagram of working principle of metallic foil electrically exploding driving flyer.

ceramics, steel or acryl glass. The base plate is used to confined the high pressure gas and plasma and reflect them to opposite direction to propel the flyers. The base plate also insulates the anode from the cathode transimission lines. So the material of base plate is non-metal and the ceramics is a good one.

The whole working process is that the large current flows through the metallic foil instantly and the metallic foil goes through from solid, to liquid, gas and plasma, and then the high pressure gases and plasmas expand to some direction to drive the polyester Mylar flyer to high velocity and impacts the targets.

Based on low inductance technologies of pulsed storaged energy capacitor, detonator switch and parallel plate transmission lines with solid films insulation, two sets of experimenta apparatuses with storaged energy of 14.4 kJ and 40 kJ were developed for launching hypervelocity flyer. The first apparatus is only consisted of one storaged energy pulsed capcitor with capacitance of 32 μF, inductance of 30 nH and rated voltage of 30 kV. The parallel plate transmission lines and solid insulation films are used, which are with very low inducatnce. The thickness of insulation films is no more than 1 mm, which is composed of several or ten pieces of Mylar films with thichness of 0.1 mm. The second apparatus is composed of two capacitors with capacitance of 16 μF and rated voltage of 50 kV in parallel For two apparatuses, the detonator switch is used, which is with low inductance of about 7 nH and easy to connected with the parallel plate transmission lines.

Fig.6 shows the diagram of the detonator switch. The detonator is exploded and the explosion products make the aluminum ring form metallic jet and breakdown the insulation films between anode and negative electrodes, and then the storaged energy is discharged to the load.

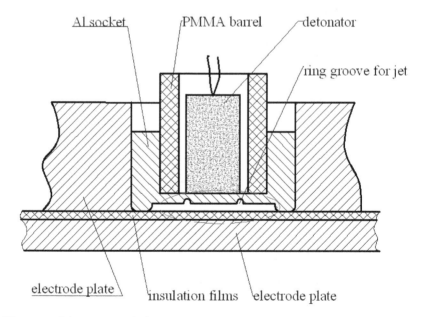

Fig. 6. Diagram of detonator switch

Fig. 7 shows the photoes of two apparatuses and Table 1 gives the electrical parameters of these two apparatuses.

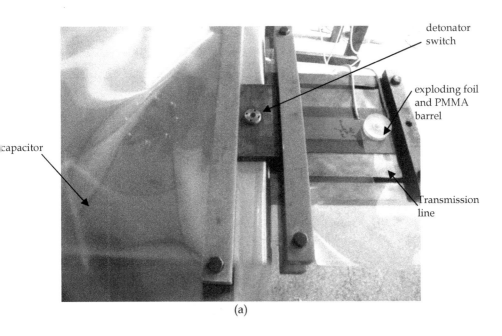

(a)

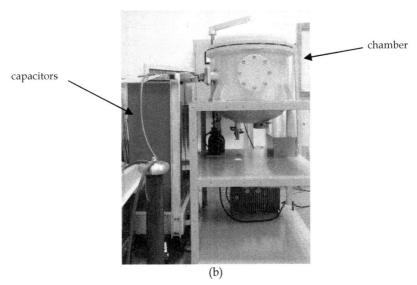

(b)

Fig. 7. Experimental apparatuses of metallic foil electrically exploding driving flyers. The apparatus with energy of 14.4 kJ (a) and the apparatus with energy of 40 kJ(b).

setup	$C/\mu F$	U_0/kV	E/kJ	$R/m\Omega$	L/nH	$T/\mu s$	$(dI/dt)_{t=0}$ $/(A/s)$	Remarks
1	32	30	14.4	14	40	7.1	7.5×10^{11}	Single capacitor
2	32	50	40	10	36	6.75	8.4×10^{11}	Two capacitors in parallel

Table 1. Parameter Values of our two apparatuses

Table 2 gives the performance parameters of our two apparatuses of metallic foil electrically exploding driving flyers.

Parameters	Setup	
	1	2
Flyer—Mylar	$\phi(6\sim20)$mmx$(0.1\sim0.2)$mm	$\phi(10\sim30)$mmx$(0.1\sim0.3)$mm
Foil—Aluminum	$(6\sim20)$mm$\times(6\sim20)$mm$\times0.028$ mm	$(10\sim30)$mm$\times(10\sim30)$mm$\times0.05$ mm
Barrel—PMMA	$\phi(6\sim20)$mmx$(4\sim15)$mm	$\phi(10\sim30)$mmx$(4\sim15)$mm
Flyer velocity	$3\sim10$km/s	$3\sim15$km/s
Flyer Simultaneity at Impact	≤25 ns	≤35 ns

Table 2. The performance parameters of our two apparatuses

The typical velocity histories of the flyers are shown in Fig.8, which are measured by laser interferometer, such as VISAR (velocity interferometer system for any reflectors)[27] or DISAR(all fibers displace interferometer system for any reflectors)[28].

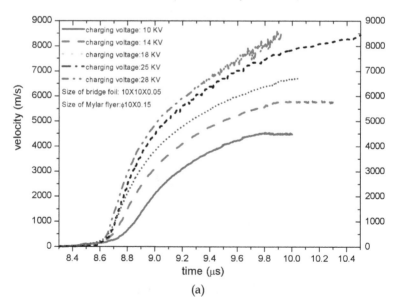

(a)

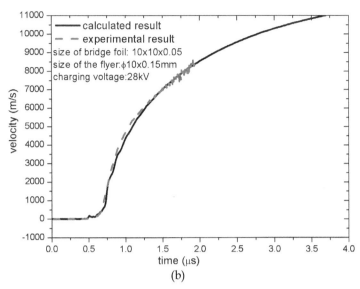

(b)

ig. 8. The experimental results of the velocity of the flyer in different conditions. The elocities of the flyers vary from charging voltages (a) and the calculated and measured elocities of the flyers (b)

As described above, the apparatus of metallic foil electrically exploding driving flyers is a ood plane wave generator for shock wave physics experiments. In the last part, we will ntroduce some important applications of this tool.

.2 Magnetically driven quasi-isentropic compression

'he techinques to realize magnetically driven quasi-isentropic compression are based on all inds of pulsed power generators, such as ZR, Veloce[29], Saturn[30] facilities. As shown in 'ig.9, Current $\vec{J}$ flowing at the anode and cathode surfaces induces a magnetic field $\vec{B}$ in

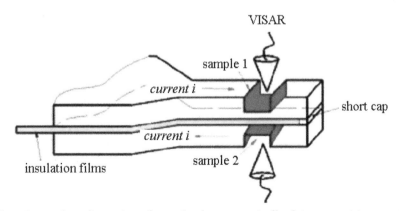

'ig. 9. Experimental configuration of samples for magnetically driven quasi-isentropic 'ompression

the gap. The resulting $\vec{J} \times \vec{B}$ Lorentz force is transferred to the electrode material, and a ramp stress wave propagates into the samples. The stress normal to the inside surfaces o electrods is $P_B = (1/2)\mu_0 J^2$, where J is the current per unit width. Two identical samples with a difference in thickness of h, are compressed by identical B-force and their particle velocity profiles $u(t)$ are measured by DISAR or VISAR.

An inverse analysis technique, i.e, the backward integration technique using difference calculation is developed to extract a compression isentrope from free-surface or window interface velocity profiles[31]. Different from Lagrangian wave analysis, inverse analysis can account for ramp-wave interactions that arise at free surfaces or window interfaces. In this method, the profiles of velocity and density are specified as an initial condition at the Lagrangian position of the measurement, then the equations of motion from equation (5) through equation (7) are integrated in the negative spatial direction to a position inside the material that is free of interaction effects during the time of interest. Assuming some parametric form shown in equation (8) for the mechanical isentrope of the material such as Murnaghan euqation or others, the parameter values are found by iteratively performing backward intergration on data from multiple thickness of the sample while minimizing the deviation between the results at a common position.

$$\sigma(h - dh, t) = \sigma(h.t) + \rho_0[u(h, t + dt) - u(h, t - dt)]dh / (2dt) \tag{5}$$

$$\varepsilon(h - dh, t) = F[\sigma(h - dh), t)] \tag{6}$$

$$u(h - dh, t) = u(h, t) + [\varepsilon(h, t + dt) - \varepsilon(h, t - dt)]dh / (2dt) \tag{7}$$

$$B_s(V) = B_{s0}\left(\frac{V_0}{V}\right)^{B'} \tag{8}$$

In order to do quasi-isentropic compression experiments, a compact capacitor bank facility CQ-1.5[13] was developed by us, which can produce a pulsed current with peak value of about 1.5 MA and rising time of 500 ns~800 ns. The solid insulating films are used to insulate the anode electrode plates from the cathode ones. And the facility is used in the air Fig.10 presents the picture of CQ-1.5.Based on CQ-1.5, about 50 GPa pressure is produced on the surface of steel samples. The parameter values of CQ-1.5 is given in Table 3.

performance parameters	values
total capacitance	15.88 μF
period in short-circuit	3.40 μs
rise time	500~800 ns
total inductance	about 18 nH
total resistance	~10 mΩ
charging voltage	75 kV~80 kV
peak current	≥1.5 MA

Table 3. The specifications of CQ-1.5

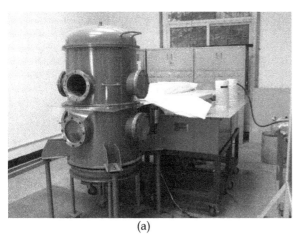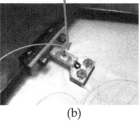

(a) (b)

Fig. 10. The picture of experimental apparatus CQ-1.5 (a) and its load area including sample and measuring probe (b).

Fig. 11. shows the typical loading pressure histories. The pressure is a ramp wave.

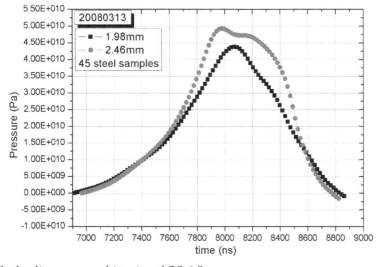

Fig. 11. The loading pressure histories of CQ-1.5

4. MHD simulation of metallic foil electrically exploding driving highvelocity flyers and magnetically driven quasi-isentropic compression

4.1 Metallic foil electrically exploding driving highvelocity flyers

The code used to simulate the electrical explosion of metallic foil is improved based our SSS code[32], which is one dimensional hydrodynamic difference code based on Lagrange orthogonal coordinate. For the case of electrical explosion of metallic foil, the power of Joule heating is increase into the energy equation, and the magnetic pressure part is considered. In order to calculate the power of Joule heating and magnetic pressure, the discharging current history is needed which is detemined by the electric circuit equation (2) and equation (3). The resistance of foil varies from different phase states during dicharging process, so a precisionly electrical resistivity model is needed to decribe this change. The physical model is seen in Figure 1, and the Lagrange hydrodynamic equations are:

$$
\begin{cases}
V = \dfrac{\partial X}{\partial M} \\[2mm]
\dfrac{\partial U}{\partial t} = -\dfrac{\partial \sigma}{\partial M} + f_{EM} \\[2mm]
\dfrac{\partial E}{\partial t} = -\dfrac{\partial(\sigma U)}{\partial M} + \dfrac{\partial(\Delta P)}{\partial M} + \lambda \dfrac{\partial}{\partial M}\left(\dfrac{\partial T}{\partial X}\right) \\[2mm]
\Delta P = I^2 R_{foil} \\[2mm]
f_{EM} = \vec{j}(X) \times \vec{B}(X) / M
\end{cases}
\tag{9}
$$

Where, V is specific volume, M is mass, X is Lagrange coordinate, U is velocity, T is temperature, λ is thermal conductivity, σ is the total pressure and $\sigma = p+q$, p is heating pressure, q is artifical viscosity pressure, f_{EM} is magnetic pressure per mass, E is total specific energy and $E = e + 0.5U^2$, e is specific internal energy, ΔP is power of Joule heating, B is magnetic flux density, μ is vacuum permeability, k is shape factor and $k=0.65$, R_{foil} is resistance of metallic foil and I is the current flowing through metallic foil in the circuit, which can be expressed with equation (10).

$$
\begin{cases}
\dfrac{1}{C_0}\displaystyle\int_0^t I(t)dt + RI + L\dfrac{dI}{dt} = Vol_0 \\[3mm]
L = L_s + L_d \\[2mm]
R = R_s + R_{foil} \\[2mm]
L_d = L(h + \Delta h) - L(h) = \dfrac{\mu l}{2\pi}\ln\dfrac{h + 1.23b}{h + \Delta h + 1.23b} \\[3mm]
R_{foil}(t) = \dfrac{l}{b}\cdot\dfrac{1}{\displaystyle\int_0^h {}^{1}\!/_{\eta(X,t)}\,dX}
\end{cases}
\tag{10}
$$

In the equation (10), C_0 is the capacitance of the experimental device, L is the total inductance of the circuit, L_s is the fixed inductance of the circuit, L_d is the variable inductance of the expansion of metallic foil caused by electrical explosion, R is the total resistance of the circuit, and R_s is the fixed resistance and R_{foil} is the dynamic resistance of the foil caused by electrical explosion, b,h and l is the width, thickness and length of the foil, η is the electrical resistivity, which is variable and can be expressed by the model put

)rward by T.J. Burgess[33]. The Burgess's model can describe the electrical resistivity of the)il at different phase states.
or solid state, there is

$$\eta_s = (C_1 + C_2 T^{C_3}) \cdot \left(\frac{V}{V_0}\right)^{F(\gamma)} \tag{11}$$

ι equation (11), C_1, C_2 and C_3 are fitting constants, γ is Gruneisen coefficient, for many ιaterials, $F(\gamma)=2\gamma-1$.
or liquid state, there is

$$\eta_L = \Delta\eta \cdot (\eta_s)_{T_m} \cdot \left(\frac{T}{T_m}\right)^{C_4} \tag{12}$$

ι equation (12), for many materials, $\Delta\eta = ke^{0.069 L_F / T_m}$, k is a constant, L_F is the melting ιtent heat, T_m is melting point temperature and C_4 is fitting constant.
or gas state, the electrical resistivity is related with both the impact between electrons and ιns and between electrons and neutrons. so,

$$\begin{cases} \eta_v = \eta_{ei} + \eta_{en} \\ \eta_{ei} = \dfrac{C_5}{T}[1 + \ln(1 + C_6 VT^{3/2})] \\ \eta = C_7 T^{1/2}[1 + \alpha_i^{-1}] \\ \partial_i = (1 + \dfrac{C_8 e^{C_9/T}}{VT^{3/2}})^{-1/2} \end{cases} \tag{13}$$

ι equation (13), α_i is the ionization fraction, C_5, C_6, C_7, C_8 and C_9 are fitting constants.
ι fact, there is mixed phase zone between liquid and gas states, a mass fraction m is ιefined. When m=0, all mass is condensed, and m=1, all mass is gas, and 0<m<1, the mass is ιixture states. Two mixture variants are also defined besides mass fraction.

$$\begin{cases} m = (V - V_0)\dfrac{C_{10}}{C_{11}} e^{-C_{12}/T} \\ X_c = (1-m)/(V/V_0) \\ X_V = 1 - X_C \end{cases} \tag{15}$$

Vhere C_{10}, C_{11} and C_{12} are fitting constants.
"he electrical resistivity of mixed phase zone can be expressed

$$\begin{cases} \eta_{mixed} = \left(\dfrac{X_C}{\eta_c} + \dfrac{X_V}{\eta_V}\right)^{-1} \\ \eta_c = \eta_l \end{cases} \tag{16}$$

"able 4 gives the parameters values of Burgess's model for Aluminum, which is used in our xperiments.

$C_1(m\Omega\text{-cm})$	C_2	C_3	C_4	C_5	C_6	γ_0	$L_F(Mbar\text{-}cm^3/mole)$
-5.35e-5	0.233	1.210	0.638	1.5	1.20e-2	2.13	0.107
C_7	C_8	C_9	C_{10}	C_{11}	C_{12}	k	$T_{m,0}$ (ev)
3.80e-3	18.5	5.96	0.440	3.58e-2	3.05	0.878	0.0804

Table 4. The parameters values of Burgess's model for Aluminum

The calculated results are presented in from Fig.12 through Fig.15. In Fig.14 and Fig.15, the experimental and calculated results are compared.

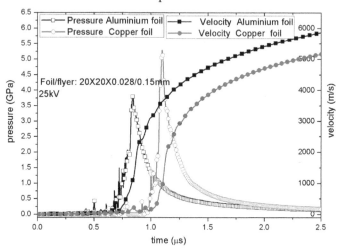

Fig. 12. The calculated pressure and flyer velocity history results of electrical explosion of Aluminum and Copper foils.

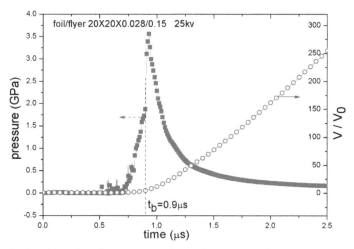

Fig. 13. The calculated results of pressure and specific volume of aluminum foil when exploding.

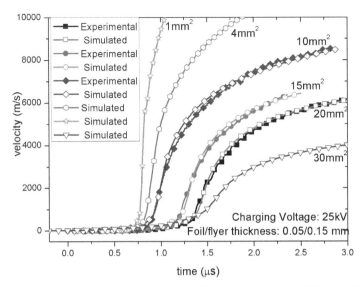

Fig. 14. The calculated and experimental results of flyer velocities for different flyer sizes.

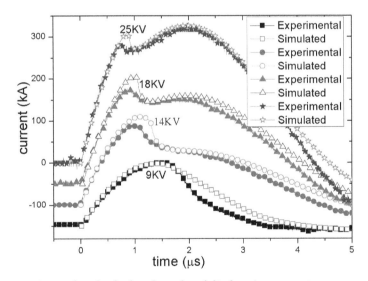

Fig. 15. The experimental and calculated results of discharging current.

The results presented in Fig.12 through Fig.15 show that the physical model here is appropriate to the electrical explosion of metallic foils.

4.2 Magnetically driven quasi-isentropic compression

In order to simplify the problem, the one dimensional model of magnetically driven quasi-isentropic compression can be described by the model shown in Fig.16. The changes of electrical parameters caused by the motion of loaded electrode are not considered, and the heat conduction is neglected because it is slow in sub microsecond or one microsecond. A standardly discharging current in short circuit is as input condition presented in Fig.17. The relative magnetic permeability is supposed tobe 1, that is to say , $\mu \cong \mu_0$.

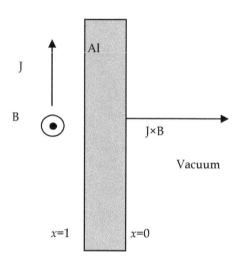

Fig. 16. Physical model of simulation

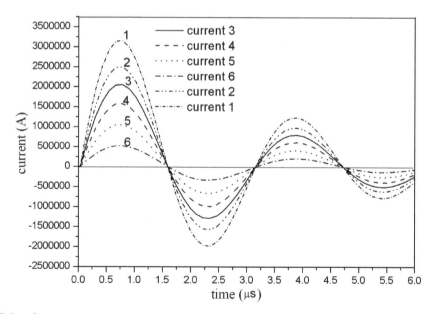

Fig. 17. Loading current curves

The controlling equations are one dimensionally magnetohydrodynamic ones, which include mass conservation equation, momentum conservation equation, energy conservation equation and magnetic diffusion equation, as shown in equation (4). The original boundary conditions are,

For $t=0$, $\begin{cases} x=0: B=0, P=0 \\ x=1: B=0, P=0 \end{cases}$, and for $t=t_n$ (at some time), $\begin{cases} x=0: B=0, P=0 \\ x=1: B=\mu_0 J(t), P=0 \end{cases}$.

The calculation coordinate are Lagrangian ones, and for the Lagrangian coordinate, the equation (4) can be converted to equations from (17) through (19).

$$\rho_0 \frac{du}{dt} + \frac{\partial}{\partial x}\left(p+q+\frac{B^2}{2\mu_0}\right) = 0 \tag{17}$$

$$\rho_0 \frac{de}{dt} + (p+q)\frac{dV}{dt} - \rho_0 \dot{e}_D = 0 \tag{18}$$

$$\frac{\partial(VB)}{\partial t} = \frac{\partial}{\partial x}\left[\frac{\eta}{\mu_0 B}\frac{\partial(VB)}{\partial x}\right] \tag{19}$$

The equation of electrical resistivity is also very important for the case of magnetically driven quasi-isentropic compression. In order to simplify the problem, a simple model is considered.

$$\eta = \eta_0 (1+\beta Q) \tag{20}$$

In equation (20), η_0 is the electrical resistivity of conductors at temperatureof 0 °C, β is heating factor, Q is the heat capacity or increment of internal energy relative to that at temperatureof 0 °C, which is related with temprature at the condensed states.

$$Q = c_v T \tag{21}$$

In equation (21), c_v is specific heat at constant volume, which is close to constant from 0 °C to the temperature of vaporazation point.

For aluminum, β is 0.69×10^{-9} m³/J, η_0 is 2.55×10^{-8} Ωm. Before vaporazation point, the equation (20) is suitable. After that, more complex electrical resisistivity model is needed. In this simulation, the stress wave front is defined when the amplitude of pressure reaches to 0.1 GPa, and thediffusion front of magnetic field is determined when the magnetic flux density is up to 0.2 T[34].

Fig.18 gives the distribution of density and temperature of Aluminum sample along Lagrangian coordinates for different times in the condition of loading current density 1.5 MA/cm.

The results in Fig.18 show that the density and temperature of aluminum sample vary with the loading time along the direction of sample thickness because of the Joule heating and magnetic field diffusion.

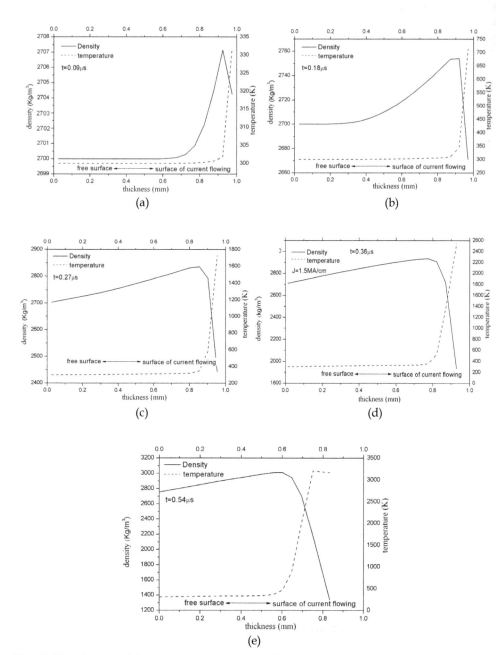

Fig. 18. Distribution of density and temperature of Aluminum sample along Lagrangian coordinates for different times under the condition of loading current density 1.5 MA/cm at time of 0.09 μs (a), 0.18 μs (b), 0.27 μs (c), 0.36 μs (d) and 0.54 μs (e)

Fig.19 gives the calculated results of distribution of magnetic induction strength along Lagrangian coordinates for different times in the condition of loading current density 1.5MA/cm.

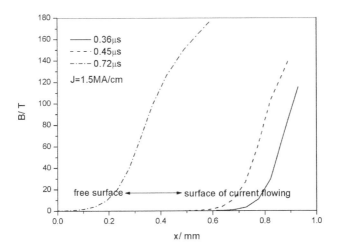

Fig. 19. Distribution of magnetic induction strength along Lagrangian coordinates for different times in the condition of loading current density 1.5MA/cm

And Fig.20 gives the physical characteristics of hydrodynamic stress wave front and magnetic diffusion front under the Lagrangian coordinates. The velocity of stress wave front is far more than that of the magnetic diffusion front, which is the prerequisite of magnetically driven quasi-isentropic compression. And the velocity of magnetic diffusion front increases gradually with the increasing of loading current density.

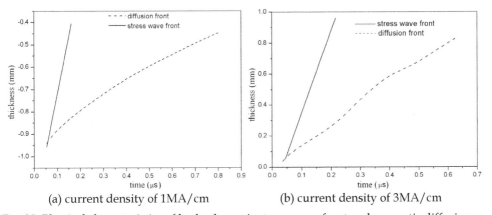

(a) current density of 1MA/cm   (b) current density of 3MA/cm

Fig. 20. Physical characteristics of hydrodynamic stress wave front and magnetic diffusion front under the Lagrangian coordinates

Fig.21 presents the relationships between the velocity of magnetic diffusion front and loading current density. The results show that an inflection poin occurs at the loading current density of 1 MA/cm, and that the results can be expressed with two linear equations (22)

$$
\begin{cases}
D = 0.008 + 0.46J, & 1.0 < J \le 3 \ \text{MA/cm} \\
\\
D = 0.36 + 0.06J, & 0.5 \le J \le 1.0 \ \text{MA/cm}
\end{cases}
\tag{22}
$$

In equation (22), D is the velocity of magnetic diffusion, and J is loading current density.

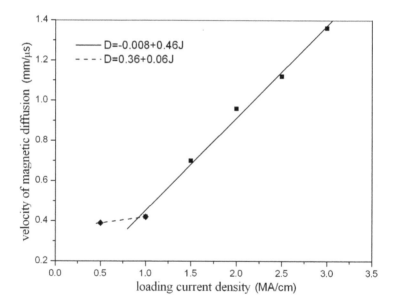

Fig. 21. The relationship of magnetic diffusion velocity varying with loading current densities.

Fig.22 is the case of copper samples under magnetically driven quasi-isentropic compression. The calculated results show that the particle velocity curves become steeper with the increasing of sample thickness, and that the shock is formed when the thickness is more than 2.5 mm for this simulating condition.

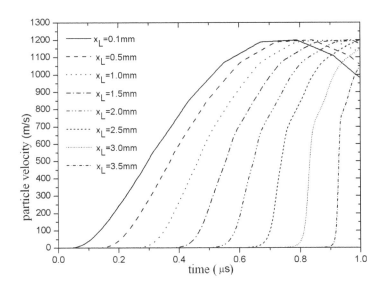

Fig. 22. The particle velocities of copper sample at different thickness in the condition of loading current density of 3 MA/cm.

5. Applications of metallic foil electrically exploding driving highvelocity flyers and magnetically driven quasi-isentropic compression

5.1 Metallic foil electrically exploding driving highvelocity flyers
5.1.1 Short-pulse shock initiation of explosive
The apparatus of metallic foil electrically exploding driving high velocity flyer offers an attractive means of performing shock initiation experiments. And the impact of an electrically exploding driven flyer produces a well-defined stimulus whose intensity and duration can be independently varied. Experiments are low-cost and there is fast turn-around between experiments.

Short-pulse shock initiation experiments will be very useful in developing more realistic theoretical shock initiation models. For the present, the models predicting shock initiation thresholds is short of, where very short pulses are employed . The technique can provide data to test the capability of improved models.

Based on our experimental apparatus, the shock initiation characteristics of TATB and TATB-based explosives are studied[35,36]. Fig.23 and Fig.24 show the experimental results of shock initiation thresholds and run distance to detonation of a TATB-based explosive.

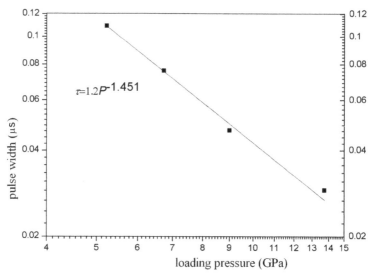

Fig. 23. Shock initiation threshold of 50% probability of initiation

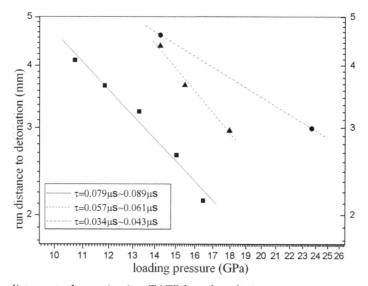

Fig. 24. Run distance to detonation in a TATB-based explosive

These experiments have the additional advantage of being applicable to relatively small explosive samples, an important consideration for evaluating and ranking new explosives.

5.1.2 Spallation experiments of materials

Compared with gas gun and explosively driven loading, The apparatus of metallic foil electrically exploding driving high velocity flyer is also a good tool used to research ynamic behaviors of materials. The loading strain rates and stress duration vary easily. In rder to study damage properties of materials using the apparatus of metallic foil lectrically exploding driving high velocity flyer, a concept of two-stage flyer is put rward[37]. The Mylar flyer flies some distance to impact a buffer plate such as PMMA or ylon with different thickness, and the pressure produced in the buffer is attenuated to the xpected value, and then the attenuated pressure propels the impactor on the buffer to some elocity to impact the target. The impactor is the same material as the target. Fig.25 is the iagram of the two-stage flyer based on the apparatus of metallic foil electrically exploding riving high velocity flyer.

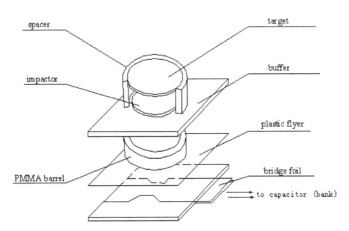

ig. 25. Sketch of two-stage flyer based on the apparatus of metallic foil electrically xploding driving high velocity flyer

ly means of the two-stage flyer, the spallations of steel and copper samples were esearched. Fig.26 is the experimental results[38].

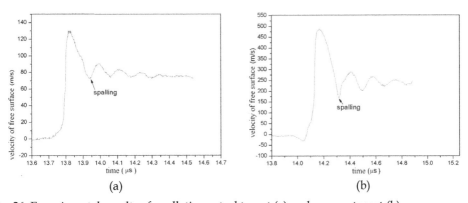

(a)   (b)

ig. 26. Experimental results of spallation , steel target (a) and copper target (b).

t is also convenient to study other dynamic behaviors of materials using the electric gun. urther experimental researches about materials are being done by our research group.

5.1.3 Potential applications

Equation of state (EOS) measurement is an important potential application for our apparatus. In order to increase the loading pressure of this apparatus, two improvements should be done. Firstly, the flyer should be Mylar-metal foil laminate flyer . The metal layer increases the flyer's shock impedance and thus the pressure produced in the target. Secondly, the storaged energy of apparatus should be increased. The expected pressure should be up to 200 GPa or more.

Impact experiment on the structure is also an important application for the apparatus of metallic foil electrically exploding driving high velocity flyer. For the apparatus of metallic foil electrically exploding driving high velocity flyer, its environment is well-controlled and instrumented, so it is suitable for studying impact phenomena in the fields of space science. Fig.27 shows a result of flyer of our apparatus impacting multi-layer structure.

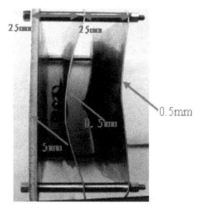

Fig. 27. Experimental result of flyer impact multi-layer structure

5.2 Magnetically driven quasi-isentropic compression

5.2.1 Compression isentropes of copper and aluminum

The experimental compression isentropes of T1 copper and L1 pure aluminum(Al content more than 99.7%) were measured on the CQ-1.5. The free-surface velocities were measured by DISAR, and the data were processed with the backward integration code developed by us. For the design of sample sizes, it is necessary that shock should not be formed in the samples and the side rarefaction wave should not affect the center regime to meet the requirements of one dimensional strain loading. Table 5 are the sizes of experimental samples.

| Exp. Shot. | materials | thickness | | width |
		$h1$	$h2$	w
20070605	T1 copper	2.00 mm	1.58 mm	7 mm
20070608-1	T1 copper	2.08 mm	1.78 mm	5 mm
20070608-2	T1 copper	2.04 mm	1.52 mm	7 mm
20070705	L1 aluminum	2.60 mm	2.02 mm	5 mm

Table 5. Experimental conditions

Fig.28(a) are the typical free-surface velocity histories measured by DISAR, which show that the slope become steeper for thicker sample. The experimental compression isentropes, theoretical compression isentropes and shock Hugoniots data are presented in Fig.28(b) and Fig.28 (c).

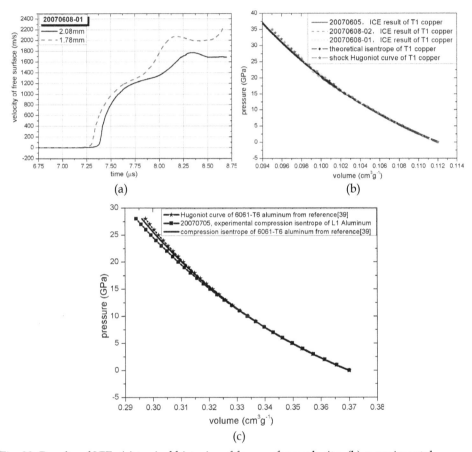

(a)

(b)

(c)

Fig. 28. Results of ICEs.(a) typical histories of free-surface velocity. (b) experimental, theoretical isentropes and Hugoniots data of T1 copper. (c) experimental isentrope of L1 pure aluminum, isentrope ang Hugoniot data of 6061-T6 aluminum from reference [39].

The results show that the experimental compression isentropes are consistent with the theoretical ones within a deviation of 3%, and are close to the shock Hugoniot data under the pressure of 40GPa and lies under them. Different from the shock method, the whole isentrope can be obtained in one shot, and tens of shots are needed to gain one shock Hugoniot curve. The calculation results[40] show that the compression isentropes gradually deviate from the shock Hugoniots with the increasement of loading pressure over 50 GPa. Therefore, the compression isentropes mainly reflect the off- Hugoniot properties of materials. Under 50 GPa, the compression isentropes are close to the shock Hugoniots, so we can use the isentrope data to check the validity or precision of shock Hugoniots.

5.2.2 Phase transition of 45 steel

Since the quasi-isentropic compression loading technique actually follows the P-v response of the material under investigation, the actual evolution of the phase trnasition can be observed. The classical polymorphic transtion of iron at 13 GPa has been studied under quasi-isentropic compression. The two free-surface velocity profiles recorded in our experiments are shown in Fig.28. The elastic precursor wave is clearly seen in the lower pressure region of the two profiles. And the plastic wave and phase change wave occur, which show that the polymorphic transition($\alpha \rightarrow \varepsilon$) takes place. The velocity profiles in Fig.29 indicates that the onset of the phase transition is at velocity of 681 m/s, and the pressure of phase transition is also about 11.4 GPa.

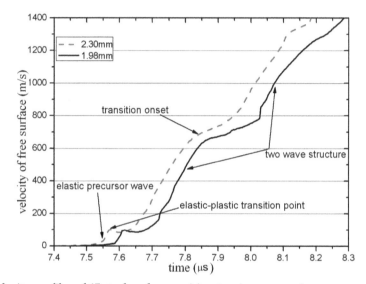

Fig. 29. Velocity profiles of 45 steel under quasi-isentropic compression

5.2.3 Spallation and elasto-plastic transition of pure tantalum

Fig.29 shows the results of the spalling experiments for pure tantalum (Ta contents 99.8%). The loading strain rate is 2.53×10^5 1/s. For the sample with thickness of 1.66 mm, the spallation is not obvious, perhaps the mirco-damage occurs. For the sample with thickness of 1.06 mm, the spallation is obvious, and the pull-back velocity is 129.6 m/s. According to the formular (23), the spall strength is 4.49 GPa.

$$\sigma_{\text{spall}} = \frac{1}{2}\rho_0 C_1 \Delta U_{\text{pb}} \qquad (23)$$

where ρ_0 is the initial density of sample, C_1 is the Larangian sound speed, ΔU_{pb} is the pull-back velocity as shown in Fig.4, and σ_{spall} is the spall strength of materials.

Under quasi-isentropic compression, the elasto-plastic transition are clearly shown in the velocity profiles of 45 steel and pure tantalum in Figure 28 and Fig.30. Here a concept of isentropic elastic limit(IEL, σ_{IEL}) is introduced. For the 45 steel sample, the σ_{IEL} is 2.26 GPa at the loading strain rate of 6.73×10^5 1/s, and for the pure tantalum sample, the σ_{IEL} is 2.42 GPa

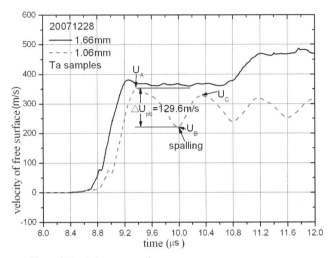

Fig. 30. Velocity profiles of Tantalum samples

at the loading strain rate of 2.53×10^5 1/s. Because of the difference of loading strain rates, the σ_{IEL} ranges from 2.26 to 2.35 GPa for 45 steel, and from 2.42 to 2.70 GPa for pure tantalum in our experiments, correspondingly, the yield strength ranges from 1.29 to 1.34 GPa for 45 steel and from 1.12 to 1.25 GPa for pure tantalum.

5.2.4 Magnetically driven high-velocity flyers

It is an important application to launch high-velocity flyer plates using the techniques of magnetically driven quasi-isentropic compression. For the present, the reseachers has launched the aluminum flyer plate with the size of 15 mm×11 mm×0.9 mm to the velocity of 43 km/s using this technique[23], and can produce 1~2 TPa shock pressure on the heavy metallic or quartz samples. Based on CQ-1.5, the aluminum flyer plate with the size of 8 mm×6 mm×0.9 mm was launched to about 9 km/s by us. Figure 31 shows the experimental results of the velocities of the flyers.

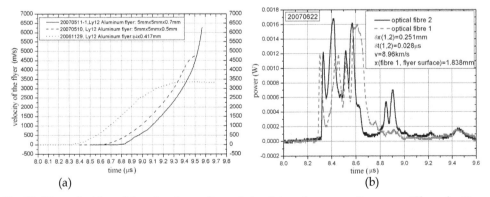

(a)   (b)

Fig. 31. The velocities of the aluminum flyer plates driven by magnetic ressure.The velocities measured by VISAR (a) and the averaged velocity measured by optical fibres pins (b)

6. Summary

The physical processes of electrical explosion of metallic foil and magnetically driven quasi-isentropic compression are very complex. This chapter dicusses these problem simply from the aspect of one dimensionally magnetohydrodyamics. The key variable of electrical resistivity was simplified, which is very improtant. Especially for the problem of magnetically driven quasi-isentropic compression, only the resistivity is considered before the vaporazation point of the matter. In fact, the phase states of the loading surface vary from solid to liquid, gas and plasma when the loading current density becomes more and more. In order to optimize the structural shapes of electrodes and the suitable sizes of samples and windows in the experiments of magnetically driven quasi-isentropic compression, two dimensionally magnetohydrodynamic simulations are necessary.

The applications of the techniques of electrical explosion of metallic foil and magnetically driven quasi-isentropic compression are various, and the word of versatile tools can be used to describe them. In this chapter, only some applications are presented. More applications are being done by us, such as the quasi-isentropic compression experiments of un-reacted solid explosives, the researches of hypervelocity impact phenomena and shock Hugoniot of materials at highly loading strain rates of $10^5 \sim 10^7$ 1/s.

7. Acknowledgements

The authors of this chapter would like to acknowledge Prof. Chengwei Sun and Dr. Fuli Tan, Ms. Jia He, Mr. Jianjun Mo and Mr. Gang Wu for the good work and assistance in our simulation and expeimental work. We would also like to express our thanks to the referee for providing invaluable and useful suggestions. Of cousre, the work is supported National Natural Science Foundation of China under Contract NO. 10927201 and NO.11002130, and the Science Foundation of CAEP under Contract NO. 2010A0201006 and NO. 2011A0101001.

8. References

[1] Keller D. V. and Penning R. J., Exploding foils – the production of plane shock waves and the acceleration of thin plates, Exploding Wires, W. G. Chase and H. K. Moore, Eds. Plenum Press, New York,Vol.2. 1962: 263

[2] Guenther A. H. ,Wunsch D. C. and Soapes T. D., Acceleration of thin plates by exploding foil techniques, Exploding Wires, W. G. Chase and H. K. Moore, Eds. Plenum Press, New York, Vol.2. 1962: 279

[3] Kotov Y A, Samatov O. M., Production of nanometer-sized AlN powders by the exploding wire method[J]. Nanostructured Materials, Vol.12(1-4),1999: 119

[4] Suzuki T, Keawchai K, Jiang W H. Nanosize Al_2O_3 powder production by pulsed wire discharge[J]. Jpn, J. Appl. Phys., Vol.40, 2001: 1073

[5] Weingart R.C. , Electric gun: applications and potential, UCRL-52000-80-2,1980

[6] Steinberg D. , Chau H., Dittbenner G. et al, The electric gun: a new method for generating shock pressure in excess of 1 TPa, UCI-17943, Sep. 1978

[7] Weingart R.C.,Chau H.H., Goosman D.R. et al, The electric gun: A new tool for ultrahigh-pressure research, UCRL-52752, April 1979

[8] Sun Chengwei, Private Communications, 2004

9] Hawke R. S., Duerre D. E., Huebel J. G. et al, Electrical Properties of Al₂O₃ under Isentropic Compression up to 500Gpa(5Mbar)[J]. J. Appl. Phys., Vol.49(6), June 1978: 3298~3303

10] Asay J. R., Isentropic Compression Experiments on the Z Accelelator. Shock Compression of Condensed Matter-1999, Edited by M. D. Furnish, L.C. Chhabildas and R. S. Hixson, 2000: 261~266

11] Avrillaud G., Courtois L., Guerre J. et al, GEPI: A Compact Pulsed Power Driver for Isentropic Compression Experiments and for Non Shocked High Velocity Flyer Plates. 14th IEEE Int·l Pulsed Power Conf., 2003: 913~916

12] Rothman S. D., Parker K. W. et al, Isentropic compression of lead and lead alloy using the Z machine, Shock Compression of Condensed Matter – 2003, 1235-1238，2004

13] Wang Guji , Sun Chengwei, Tan Fuli et al, The compact capacitor bank CQ-1.5 employed in magnetically drivenisentropic compression and high velocity flyer plate experiments, REVIEW OF SCIENTIFIC INSTRUMENTS 79, 053904 ,2008

14] Lemke R. W., Knudson M. D., Bliss D. E. et al, Magnetically accelerated, ultrahigh velocity flyer plates for shock wave experiments, J. Appl. Phys. 98, 2005:073530-1~9

15] Wang Guiji, Zhao Jianheng, Tang Xiaosong et al, Study on the technique of electric gun loading for one dimensionally planar strain, Chinese Journal of High Pressure Physics, Vol.19(3), 2005: 269-274

16] Brechov· Vladimir· Anatonievich, Electrical explosion of conductors and its applications in electrically physical facilities(in Russian), 2000

17] Chau H.H., Dittbenner G., Hofer W.W. et al, Electric gun: a versatile tool for high-pressure shock wave research, Rev. Sci. Instrum. 51(12), Dec. 1980, P1676~1681

18] Tucker T.J. , Stanton P.L. , Electrical gurney energy: A new concept in modeling of energy transfer from electrically exploded conductors, SAND-75-0244, May 1975

19] Schmidt S.C., Seitz W.L., Wackerle Jerry, An empirical model to compute the velocity history of flyers driven by electrically exploding foils, LA-6809, July 1977

20] He Jia, Simulation on dynamic process of metallic foil electrical explosion driving multi-stage flyers, paper for Master degree, Institute of Fluid Physics, China Academy of Engineering Physics, Mianyang, Sichuan, China, 2007

21] Asay J.R. and Knudson M.D., Use of pulsed magnetic fields for quasi-isentropic compression experiments, High-Pressure Shock Compression Solids VIII, edited by L.C. Chhabildas, L. Davison and Y. Horie, Springer,2005:329

22] Davis J. P., Deeney C., Knudson M. D. et al, Magnetically driven isentropic compression to multimegabar pressures using shaped current pulses on the Z accelerator[J]. Physics of Plasma, 12, 2005:056310-1~056310-7

23] Savage Mark , The Z pulsed power driver since refurbishment,The 13th International Conference on Megagauss Magnetic Field Generation and Related Topics, July 2010.

24] Zhao Jianheng, Sun Chengwei, Tang Xiaosong et al, The Development of high performance electric gun facility, Experimental Mechanics, Vol.21(3), 2006

25] Wang Guiji, He Jia, Zhao Jianheng et al, The Techniques of Metallic Foil Electrically Exploding Driving Hypervelocity Flyer to more than 10km/s for Shock Wave Physics Experiments, submitted to Rev. Scie. Instrum., 2011

[26] Wang Guiji, Deng Xiangyang, Tan Fuli et al, Velocity measurement of the small size flyer of an exploding foil initiator, Explosion and Shock Waves, Vol.28(1),2008 : 28-32

[27] Barker L.M. and Hollenback R.E., Laser interferometer for measuring high velocity of any reflecting surface. J. Appl. Phys., Vol. 43(11),1972: 4669~4675

[28] Weng Jidong , Tan Hua, Wang Xiang et al, Optical-fiber interferometer for velocity measurements with picosecond resolution, Appl. Phys. Lett. 89, 111101,2006

[29] Ao T. , Asay J.R., Chantrenne S. et al., A compact strip-line pulsed power generator for isentropic compression experiments, Rev. Scie. Instrum., 79(1), 013901, 2008

[30] Furnish Michael D., Davis Jean-Paul, Knudson Marcus et al, Using the Saturn Accelerator for Isentropic Compression Experiments (ICE), SAND2001-3773, Sandia National Laboratories, 2001

[31] Hayes D., Backward integration of the equations of motion to correct for free surface perturbaritz, SAND2001-1440, Sandia National Laboratories, 2001

[32] Sun Chengwei, One dimensional shock and detonation wave computation code SSS, Computation Physics, No.3, 1986: 143-145

[33] Burgess T.J., Electrical resistivity model of metals, 1986

[34] Lemke R.W., Knudson M.D. et al., Characterization of magnetically accelerated flyer plates, Phys. Plasmas 10 (4), 1092-1099, 2003

[35] Wang Guiji, Zhao Tonghu, Mo Jianjun et al., Short-duration pulse shock initiation characteristics of a TATB/HMX-based polymer bonded explosive, Explosion and Shock Waves, Vol.27(3), 2007:230-235

[36] Wang Guiji, Zhao Tonghu, Mo Jianjun et al., Run distance to detonation in a TATB/HMX-based explosive, Explosion and Shock Waves, Vol.26(6), 2006:510-515

[37] Sun Chengwei, Dynamic micro-fracture of metals under shock loading by electric gun, J. Phys.IV, Vol.4(8),1994:355-360

[38] Xiong Xin, The spallation of ductile metals under loading of electric gun driven metallic flyer, paper for Master degree, Institute of Fluid Physics, China Academy of Engineering Physics, Mianyang, Sichuan, China, 2007

[39] Hayes D. B., Hall C. A., Asay J. R. et al, Measurement of the Compression Isentrope for 6061-T6 Aluminum to 185 GPa and 46% Volumetric Strain Using Pulsed Magnetic Loading. J. Appl. Phys., Vol.96(10),2004:5520~5527

[40] Wang Ganghua, Experiments, simulation and data processing methods of magnetically driven isentropic compression and highvelocity flyer plates, paper for Ph.D degree, Institute of Fluid Physics, China Academy of Engineering Physics, Mianyang, Sichuan, China, 2008

Part 2

Special Topics on Simulations and Experimental Data

5

Elasto-Hydrodynamics of Quasicrystals and Its Applications

Tian You Fan[1] and Zhi Yi Tang[2]
[1]Department of Physics, Beijing Institute of Technology, Beijing
[2]Southwest Jiaotong University Hope College, Nanchong, Sichuan
China

1. Introduction

Quasicrystal as a new structure of solids as well as a new material, has been studied over twenty five years. The elasticity and defects play a central role in field of mechanical behaviour of the material, see e.g. Fan [1]. Different from crystals and conventional engineering materials, quasicrystals have two different displacement fields: phonon field $u(u_1,u_2,u_3)$ and phason field $w(w_1,w_2,w_3)$, which is a new degree of freedom to condensed matter physics as well as continuum mechanics, this leads to two strain tensors such as

$$\varepsilon_{ij} = \frac{1}{2}(\frac{\partial u_i}{\partial x_j} + \frac{\partial u_j}{\partial x_i}) \;,\; w_{ij} = \frac{\partial w_i}{\partial x_j} \qquad (1)$$

We call the first of equation (1) as phonon strain tensor, the second as phason strain tensor, respectively. The corresponding stress tensor is σ_{ij} and H_{ij}.

The constitutive law is the so-called generalized Hooke's law as follows

$$\sigma_{ij} = C_{ijkl}\varepsilon_{kl} + R_{ijkl}w_{kl}$$
$$H_{ij} = K_{ijkl}w_{kl} + R_{klij}\varepsilon_{kl} \qquad (2)$$

in which C_{ijkl} denotes the phonon elastic tensor, K_{ijkl} the phason one, and R_{ijkl} the phonon-phason coupling one, respectively. It is evident that the appearance of the new degree freedom yields a great challenge to the continuum mechanics.

In the dynamic process of quasicrystals problem presents further complexity. According to the point of view of Lubensky et al. [2,3], phonon represents wave propagation, while phason represents diffusion in the dynamic process. Following the argument of Lubensky et al., Rochal and Lorman [4] and Fan [1,5] put forward the equations of motion of quasicrystals as follows

$$\rho\frac{\partial^2 u_i}{\partial t^2} = \frac{\partial \sigma_{ij}}{\partial x_j} \qquad (3)$$

$$\kappa \frac{\partial w_i}{\partial t} = \frac{\partial H_{ij}}{\partial x_j} \tag{4}$$

Equation (3) is the equation of motion of conventional elastodynamics, and equation (4) is the linearized equation of hydrodynamics of Lubensky et al., so equations (3), (4) are elasto-hydrodynamic equations of quasicrystals.

The equations (1)-(4) are the basis of dynamic analysis of quasicrystalline material.

2. The elasto-hydrodynamics of two-dimensional decagonal quasicrystals and application to dynamic fracture

2.1 Statement of formulation and sample problem

Among over 200 quasicrystals observed to date, there are over 70 two-dimensional decagonal quasicrystals, so this kind of solid phases play an important role in the material. For simplicity, here only point group 10mm two-dimensional decagonal quasicrystals will be considered. We denote the periodic direction as the z axis and the quasiperiodic plane as the $x-y$ plane. Assume that a Griffith crack in the solid along the periodic direction, i.e., the z axis. It is obvious that elastic field induced by a uniform tensile stress at upper and lower surfaces of the specimen is independent of z, so $\partial(\ )/\partial z = 0$. In this case, the stress-strain relations are reduced to

$$
\begin{aligned}
\sigma_{xx} &= L(\varepsilon_{xx} + \varepsilon_{yy}) + 2M\varepsilon_{xx} + R(w_{xx} + w_{yy}) \\
\sigma_{yy} &= L(\varepsilon_{xx} + \varepsilon_{yy}) + 2M\varepsilon_{yy} - R(w_{xx} + w_{yy}) \\
\sigma_{xy} &= \sigma_{yx} = 2M\varepsilon_{xy} + R(w_{yx} - w_{xy}) \\
H_{xx} &= K_1 w_{xx} + K_2 w_{yy} + R(\varepsilon_{xx} - \varepsilon_{yy}) \\
H_{yy} &= K_1 w_{yy} + K_2 w_{xx} + R(\varepsilon_{xx} - \varepsilon_{yy}) \\
H_{xy} &= K_1 w_{xy} - K_2 w_{yx} - 2R\varepsilon_{xy} \\
H_{yx} &= K_1 w_{yx} - K_2 w_{xy} + 2R\varepsilon_{xy}
\end{aligned}
\tag{5}
$$

where $L = C_{12}, M = (C_{11} - C_{12})/2$ are the phonon elastic constants, K_1 and K_2 are the phason elastic constants, R phonon-phason coupling elastic constant, respectively.

Substituting equations (5) into equations (3), (4) we obtain the equations of motion of decagonal quasicrystals as following:

$$
\begin{aligned}
\frac{\partial^2 u_x}{\partial t^2} &= c_1^2 \frac{\partial^2 u_x}{\partial x^2} + (c_1^2 - c_2^2)\frac{\partial^2 u_y}{\partial x \partial y} + c_2^2 \frac{\partial^2 u_x}{\partial y^2} + c_3^2 (\frac{\partial^2 w_x}{\partial x^2} + 2\frac{\partial^2 w_y}{\partial x \partial y} - \frac{\partial^2 w_x}{\partial y^2}) \\
\frac{\partial^2 u_y}{\partial t^2} &= c_2^2 \frac{\partial^2 u_y}{\partial x^2} + (c_1^2 - c_2^2)\frac{\partial^2 u_x}{\partial x \partial y} + c_1^2 \frac{\partial^2 u_y}{\partial y^2} + c_3^2 (\frac{\partial^2 w_y}{\partial x^2} - 2\frac{\partial^2 w_x}{\partial x \partial y} - \frac{\partial^2 w_y}{\partial y^2}) \\
\frac{\partial w_x}{\partial t} &= d_1^2 (\frac{\partial^2 w_x}{\partial x^2} + \frac{\partial^2 w_x}{\partial y^2}) + d_2^2 (\frac{\partial^2 u_x}{\partial x^2} - 2\frac{\partial^2 u_y}{\partial x \partial y} - \frac{\partial^2 u_x}{\partial y^2}) \\
\frac{\partial w_y}{\partial t} &= d_1^2 (\frac{\partial^2 w_y}{\partial x^2} + \frac{\partial^2 w_y}{\partial y^2}) + d_2^2 (\frac{\partial^2 u_y}{\partial x^2} + 2\frac{\partial^2 u_x}{\partial x \partial y} - \frac{\partial^2 u_y}{\partial y^2})
\end{aligned}
\tag{6}
$$

where

$$c_1 = \sqrt{\frac{L+2M}{\rho}}, c_2 = \sqrt{\frac{M}{\rho}}, c_3 = \sqrt{\frac{R}{\rho}}, d_1 = \sqrt{\frac{K_1}{\kappa}}, d_2 = \sqrt{\frac{R}{\kappa}}$$

Note that constants c_1, c_2 and c_3 have the meaning of elastic wave speeds, while d_1 and d_2 do not represent wave speed, and d_1^2 and d_2^2 are diffusive coefficients in physical meaning. A decagonal quasicrystal with a crack is shown in Fig.1. It is a rectangular specimen with a central crack of length $2a(t)$ subjected to a dynamic or static tensile stress at its edges ED and FC, in which $a(t)$ represents the crack length being a function of time, and for dynamic initiation of crack growth, the crack is stable, so $a(t) = a_0 = $ constant, for fast crack propagation, $a(t)$ varies with time. At first we consider dynamic initiation of crack growth, then study crack fast propagation. Due to the symmetry of the specimen only the upper right quarter is considered.

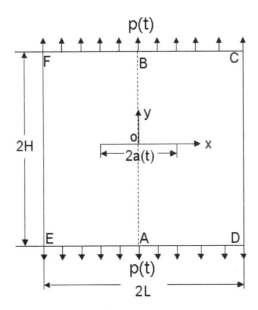

Fig. 1. The specimen with a central crack

Referring to the upper right part and considering a fix grips case, the following boundary conditions should be satisfied:

$$\begin{aligned}
&u_x = 0, \sigma_{yx} = 0, w_x = 0, H_{yx} = 0 && \text{on } x = 0 \text{ for } 0 \le y \le H \\
&\sigma_{xx} = 0, \sigma_{yx} = 0, H_{xx} = 0, H_{yx} = 0 && \text{on } x = L \text{ for } 0 \le y \le H \\
&\sigma_{yy} = p(t), \sigma_{xy} = 0, H_{yy} = 0, H_{xy} = 0 && \text{on } y = H \text{ for } 0 \le x \le L \\
&\sigma_{yy} = 0, \sigma_{xy} = 0, H_{yy} = 0, H_{xy} = 0 && \text{on } y = 0 \text{ for } 0 \le x \le a(t) \\
&u_y = 0, \sigma_{xy} = 0, w_y = 0, H_{xy} = 0 && \text{on } y = 0 \text{ for } a(t) \le x \le L
\end{aligned} \qquad (7)$$

in which $p(t) = p_0 f(t)$ is a dynamic load if $f(t)$ varies with time, otherwise it is a static load (i.e., if $f(t) = const$), and $p_0 = const$ with the stress dimension. .
The initial conditions are

$$
\begin{aligned}
& u_x(x,y,t)\big|_{t=0} = 0 \qquad u_y(x,y,t)\big|_{t=0} = 0 \\
& w_x(x,y,t)\big|_{t=0} = 0 \qquad w_y(x,y,t)\big|_{t=0} = 0 \\
& \frac{\partial u_x(x,y,t)}{\partial t}\big|_{t=0} = 0 \qquad \frac{\partial u_y(x,y,t)}{\partial t}\big|_{t=0} = 0
\end{aligned}
$$

(8

For implementation of finite difference all field variables in governing equations (6) and boundary-initial conditions (7), (8) must be expressed by displacements and their derivatives. This can be done through the constitutive equations (2). The detail of the finite difference scheme is omitted here but can be referred to Fan [1].
For the related parameters in this section, the experimentally determined mass density for decagonal Al-Ni-Co quasicrystal $\rho = 4.186 \times 10^{-3} \text{g} \cdot \text{mm}^{-3}$ is used and phonon elastic moduli are $C_{11} = 2.3433 \times 10^{12} \text{dyn/cm}^2 , C_{12} = 0.5741 \times 10^{12} \text{dyn/cm}^2$ $(10^{10} \text{dyn/cm}^2 = \text{GPa})$ which are obtained by resonant ultrasound spectroscopy, refer to Chernikov et al [6], we have also chosen phason elastic constants $K_1 = 1.22 \times 10^{12} \text{dyn/cm}^2$ and $K_2 = 0.24 \times 10^{12} \text{dyn/cm}^2$ $(10^{10} \text{dyn/cm}^2 = \text{GPa})$ estimated by Monto-Carlo simulation given by Jeong and Steinhardt [7] and $\Gamma_w = 1/\kappa = 4.8 \times 10^{-19} \text{m}^3 \cdot \text{s/kg} = 4.8 \times 10^{-10} \text{cm}^3 \cdot \mu\text{s}/\text{g}$ which measured by de Boussieu and collected by Walz in his master thesis [8]. The coupling constant R has been measured for some special cases recently, see Chapter 6 and Chapter 9 of monograph written by Fan [1] respectively. In computation we take $R/M = 0.01$ for coupling case corresponding to quasicrystals, and $R/M = 0$ for decoupled case which corresponds to crystals.

2.2 Examination on the physical model
In order to verify the correctness of the suggested model and the numerical simulation, we first explore the specimen without a crack. We know that there are the fundamental solutions characterizing time variation natures based on wave propagation of phonon field and on motion of diffusion of phason, respectively according to mathematical physics

$$
\begin{cases}
u \sim e^{i\omega(t-x/c)} \\
w \sim \dfrac{1}{\sqrt{t-t_0}} e^{-(x-x_0)^2/\Gamma_w(t-t_0)}
\end{cases}
$$

(9)

where ω is a frequency and c a speed of the wave, t the time and t_0 a special value of t, x the distance, x_0 a special value of x, and Γ_w the kinetic coefficient of phason defined previously.
Comparison results are shown in Fig.2 (a-c), in which the solid line represents the numerical solution of quasicrystals and the dotted line represents fundamental solution given by formulas (9). From Fig. 2(a) and (b) we can see that both displacement components of phonon field are in excellent agreement to the fundament solutions of mathematical physics. However, there are some differences because the phonon field is influenced by phason field

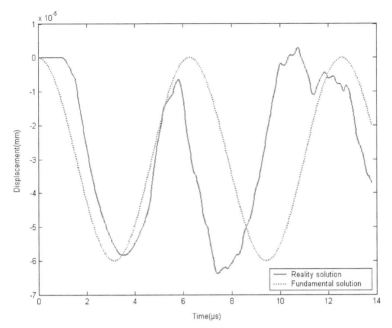

Fig. 2. (a) Displacement component of phonon field u_x versus time

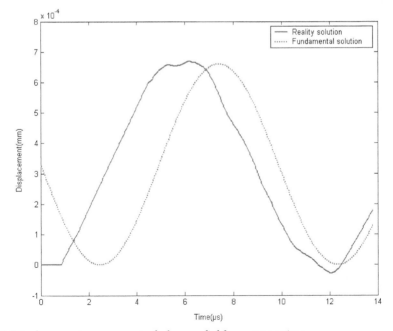

Fig. 2. (b) Displacement component of phonon field u_y versus time

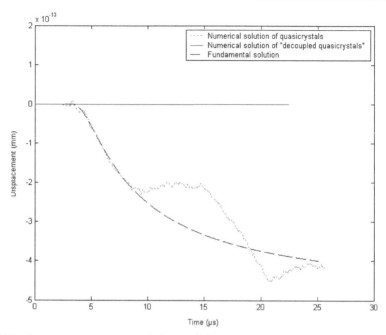

Fig. 2. (c) Displacement component of phason field w_x versus time

and the phonon-phason coupling effect. From Fig. 2(c), in the phason field we find that the phason mode presents diffusive nature in the overall tendency, but because of influence of the phonon and phonon-phason coupling, it can also have some characters of fluctuation. So the model describes the dynamic behaviour of phonon field and phason field in deed. This also shows the mathematical modeling of the present work is valid.

2.3 Testing the scheme and the computer program
2.3.1 Stability of the scheme
The stability of the scheme is the core problem of finite difference method which depends upon the choice of parameter $\alpha = c_1\tau / h$, which is the ratio between time step and space step substantively. The choice is related to the ratio c_1 / c_2, i.e., the ratio between speeds of elastic longitudinal and transverse waves of the phonon field. To determine the upper bound for the ration to guarantee the stability, according to our computational practice and considering the experiences of computations for conventional materials, we choose $\alpha = 0.8$ in all cases and results are stable.

2.3.2 Accuracy test
The stability is only a necessary condition for successful computation. We must check the accuracy of the numerical solution. This can be realized through some comparison with some well-known classical solutions (analytic as well as numerical solutions) of conventional fracture mechanics. For this purpose the material constants in the computation are chosen as $c_1 = 7.34 \, \text{mm}/\mu\text{s}, c_2 = 3.92 \, \text{mm}/\mu\text{s}$ and $\rho = 5 \times 10^3 \, \text{kg}/\text{m}^3$, $p_0 = 1 \, \text{MPa}$ which

are the same with those given in classical references for conventional fracture dynamics, discussed in Fan's monograph [1] in detail. At first the comparison to the classical exact analytic solution is carried out, in this case we put $w_x = w_y = 0$ (i.e., $K_1 = K_2 = R = 0$) for the numerical solution. The comparison has been done with the key physical quantity— dynamic stress intensity factor, which is defined by

$$K_1(t) = \lim_{x \to a_0^+} \sqrt{\pi(x - a_0)} \sigma_{yy}(x, 0, t) \tag{10}$$

The normalized dynamic stress intensity factor can be denoted as $K_I(t) / K_I^{static}$, in which K_I^{static} is the corresponding static stress intensity factor, whose value here is taken as $\sqrt{\pi a_0} p_0$. For the dynamic initiation of crack growth in classical fracture dynamics there is the only exact analytic solution— the Maue's solution (refer to Fan's monograph [1]), but the configuration of whose specimen is quite different from that of our specimen. Maue studied a semi-infinite crack in an infinite body, and subjected to a Heaviside impact loading at the crack surface. While our specimen is a finite size rectangular plate with a central crack, and the applied stress is at the external boundary of the specimen. Generally the Maue's model cannot describe the interaction between wave and external boundary. However, consider a very short time interval, i.e., during the period between the stress wave from the external boundary arriving at the crack tip (this time is denoted by t_1) and before the reflecting by external boundary stress wave emanating from the crack tip in the finite size specimen (the time is marked as t_2). During this special very short time interval our specimen can be seen as an "infinite specimen". The comparison given by Fig. 3 shows the numerical results are in excellent agreement with those of Maue's solution within the short interval in which the solution is valid. Our solution corresponding to case of $w_x = w_y = 0$ is also compared with numerical solutions of conventional crystals, e.g. Murti's solution and Chen's solutions (refer to Fan [1] and Zhu and Fan [9] for the detail), which are also shown in Fig. 3, it is evident, our solution presents very high precise.

2.3.3 Influence of mesh size (space step)

The mesh size or the space step of the algorithm can influence the computational accuracy too. To check the accuracy of the algorithm we take different space steps shown in Table 1, which indicates if $h = a_0 / 40$ the accuracy is good enough. The check is carried out through static solution, because the static crack problem in infinite body of decagonal quasicrystals has exact solution given in Chapter 8 of monograph given by Fan [1], and the normalized static intensity factor is equal to unit. In the static case, there is no wave propagation effect, $L / a_0 \geq 3, H / a_0 \geq 3$ the effect of boundary to solution is very weak, and for our present specimen $L / a_0 \geq 4, H / a_0 \geq 8$, which may be seen as an infinite specimen, so the normalized static stress intensity factor is approximately but with highly precise equal to unit. The table shows that the algorithm is with a quite highly accuracy when $h = a_0 / 40$.

2.4 Results of dynamic initiation of crack growth

The dynamic crack problem presents two "phases" in the process: the dynamic initiation of crack growth and fast crack propagation. In the phase of dynamic initiation of crack growth, the length of the crack is constant, assuming $a(t) = a_0$. The specimen with stationary crack

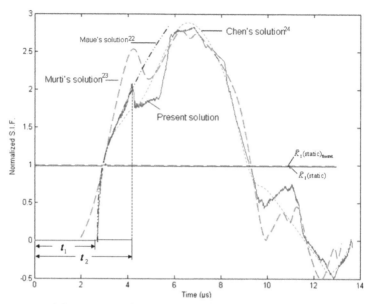

Fig. 3. Comparison of the present solution with analytic solution and other numerical solution for conventional structural materials given by other authors

H	$a_0/10$	$a_0/15$	$a_0/20$	$a_0/30$	$a_0/40$
K	0.9259	0.94829	0.9229	0.97723	0.99516
Errors	7.410%	5.171%	3.771%	2.277%	0.484%

Table 1. The normalized static S.I.F. of quasicrystals for different space steps

that are subjected to a rapidly varying applied load $p(t) = p_0 f(t)$, where p_0 is a constant with stress dimension and $f(t)$ is taken as the Heaviside function. It is well known the coupling effect between phonon and phason is very important, which reveals the distinctive physical properties including mechanical properties, and makes quasicrystals distinguish the periodic crystals. So studying the coupling effect is significant.

The dynamic stress intensity factor $K_1(t)$ for quasicrystals has the same definition given by equation (10), whose numerical results are plotted in Fig. 4, where the normalized dynamics stress intensity factor $K_1(t) / \sqrt{\pi a_0} p_0$ is used. There are two curves in the Fig. 4, one represents quasicrystal, i.e., $R / M = 0.01$, the other describes periodic crystals corresponding to $R / M = 0$, the two curves of the Fig. 4 are apparently different, though they are similar to some extends. Because of the phonon-phason coupling effect, the mechanical properties of the quasicrystals are obviously different from the classical crystals. Thus, the coupling effect plays an important role.

In Fig. 4, t_0 represents the time that the wave from the external boundary propagates to the crack surface, in which $t_0 = 2.6735 \, \mu s$. So the velocity of the wave propagation is $v_0 = H / t_0 = 7.4807 \, km/s$, which is just equal to the longitudinal wave speed $c_1 = \sqrt{(L + 2M) / \rho}$. This indicates that for the complex system of wave propagation-motion of diffusion coupling, the phonon wave propagation presents dominating role.

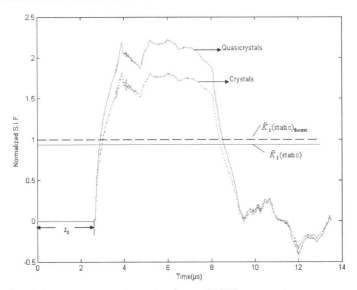

Fig. 4. Normalized dynamics stress intensity factor (DSIF) versus time

There are some oscillations of values of the stress intensity factor in the figure. These oscillations characterize the reflection and diffraction between waves coming from the crack surface and the specimen boundary surfaces. The oscillations are influenced by the material constants and specimen geometry including the shape and size very much.

3. Elasto-/hydro-dynamics and applications to fracture dynamics of three-dimensional icosahedral quasicrystals

3.1 Basic equations, boundary and initial conditions

There are over 50% icosahedral quasicrystals among observed the quasicrystals to date, this shows this kind of systems in the material presents the most importance. Within icosahedral quasicrystals, the icosahedral Al-Pd-Mn quasicrystals are concerned in particular by researchers, for which especially a rich set of experimental data for elastic constants accumulated so far, this is useful to the computational practice. So we focus on the elasto-hydrodynamics of icosahedral Al-Pd-Mn quasicrystals here. From the previous section we have known there are lack of measured data for phason elastic constants, the computation has to take some data which are obtained by Monte Carlo simulation, this makes some undetermined factors for computational results for decagonal quasicrystals. This shows the discussion on icosahedral quasicrystals is more necessary, and the formalism and numerical results are presented in the following.

If considering only the plane problem, especially for the crack problems, there are much of similarities with those discussed in the previous section. We present herein only the part that are different.

For the plane problem, i.e.,

$$\frac{\partial(\ )}{\partial z} = 0 \tag{11}$$

The linearized elasto-hydrodynamics of icosahedral quasicrystals have non-zero displacements u_z, w_z apart from u_x, u_y, w_x, w_y, so in the strain tensors

$$\varepsilon_{ij} = \frac{1}{2}\left(\frac{\partial u_i}{\partial x_j} + \frac{\partial u_j}{\partial x_i}\right) \qquad w_{ij} = \frac{\partial w_i}{\partial x_j}$$

it increases some non-zero components compared with those in two-dimensional quasicrystals. In connecting with this, in the stress tensors, the non-zero components increase too relatively to two-dimensional ones. With these reasons, the stress-strain relation presents different nature with that of decagonal quasicrystals though the generalized Hooke's law has the same form with that in one- and two-dimensional quasicrystals, i.e.,

$$\sigma_{ij} = C_{ijkl}\varepsilon_{kl} + R_{ijkl}w_{kl} \qquad H_{ij} = R_{klij}\varepsilon_{kl} + K_{ijkl}w_{kl}$$

In particular the elastic constants are quite different from those discussed in the previous sections, in which the phonon elastic constants can be expressed such as

$$C_{ijkl} = \lambda\delta_{ij}\delta_{kl} + \mu(\delta_{ik}\delta_{jl} + \delta_{il}\delta_{jk}) \tag{12}$$

and the phason elastic constant matrix [K] and phonon-phason coupling elastic one [R] are defined by the formulas of Fan's monograph [1], which are not listed here again. Substituting these non-zero stress components into the equations of motion

$$\rho\frac{\partial^2 u_i}{\partial t^2} = \frac{\partial \sigma_{ij}}{\partial x_j}, \; \kappa\frac{\partial w_i}{\partial t} = \frac{\partial H_{ij}}{\partial x_j} \tag{13}$$

and through the generalized Hooke's law and strain-displacement relation we obtain the final dynamic equations as follows

$$\frac{\partial^2 u_x}{\partial t^2} + \theta\frac{\partial u_x}{\partial t} = c_1^2\frac{\partial^2 u_x}{\partial x^2} + (c_1^2 - c_2^2)\frac{\partial^2 u_y}{\partial x\partial y} + c_2^2\frac{\partial^2 u_x}{\partial y^2} + c_3^2\left(\frac{\partial^2 w_x}{\partial x^2} + 2\frac{\partial^2 w_y}{\partial x\partial y} - \frac{\partial^2 w_x}{\partial y^2}\right)$$

$$\frac{\partial^2 u_y}{\partial t^2} + \theta\frac{\partial u_y}{\partial t} = c_2^2\frac{\partial^2 u_y}{\partial x^2} + (c_1^2 - c_2^2)\frac{\partial^2 u_x}{\partial x\partial y} + c_1^2\frac{\partial^2 u_y}{\partial y^2} + c_3^2\left(\frac{\partial^2 w_y}{\partial x^2} - 2\frac{\partial^2 w_x}{\partial x\partial y} - \frac{\partial^2 w_y}{\partial y^2}\right)$$

$$\frac{\partial^2 u_z}{\partial t^2} + \theta\frac{\partial u_z}{\partial t} = c_2^2\left(\frac{\partial^2}{\partial x^2} + \frac{\partial^2}{\partial y^2}\right)u_z + c_3^2\left(\frac{\partial^2 w_x}{\partial x^2} - \frac{\partial^2 w_x}{\partial y^2} - 2\frac{\partial^2 w_y}{\partial x\partial y} + \frac{\partial^2 w_z}{\partial x^2} + \frac{\partial^2 w_z}{\partial y^2}\right)$$

$$\frac{\partial w_x}{\partial t} + \theta w_x = d_1\left(\frac{\partial^2}{\partial x^2} + \frac{\partial^2}{\partial y^2}\right)w_x + d_2\left(\frac{\partial^2}{\partial x^2} - \frac{\partial^2}{\partial y^2}\right)w_z + d_3\left(\frac{\partial^2 u_x}{\partial x^2} - 2\frac{\partial^2 u_y}{\partial x\partial y} - \frac{\partial^2 u_x}{\partial y^2} + \frac{\partial^2 u_z}{\partial x^2} - \frac{\partial^2 u_z}{\partial y^2}\right)$$

$$\frac{\partial w_y}{\partial t} + \theta w_y = d_1\left(\frac{\partial^2}{\partial x^2} + \frac{\partial^2}{\partial y^2}\right)w_y - d_2\frac{\partial^2 w_z}{\partial x\partial y} + d_3\left(\frac{\partial^2 u_y}{\partial x^2} + 2\frac{\partial^2 u_x}{\partial x\partial y} - \frac{\partial^2 u_y}{\partial y^2} - 2\frac{\partial^2 u_z}{\partial x\partial y}\right)$$

$$\frac{\partial w_z}{\partial t} + \theta w_z = (d_1 - d_2)\left(\frac{\partial^2}{\partial x^2} + \frac{\partial^2}{\partial y^2}\right)w_z + d_2\left(\frac{\partial^2 w_x}{\partial x^2} - \frac{\partial^2 w_x}{\partial y^2} - 2\frac{\partial^2 w_y}{\partial x\partial y}\right) + d_3\left(\frac{\partial^2}{\partial x^2} + \frac{\partial^2}{\partial y^2}\right)u_z \tag{14}$$

which

$$c_1 = \sqrt{\frac{\lambda + 2\mu}{\rho}}, c_2 = \sqrt{\frac{\mu}{\rho}}, c_3 = \sqrt{\frac{R}{\rho}}, d_1 = \frac{K_1}{\kappa}, \quad d_2 = \frac{K_2}{\kappa}, d_3 = \frac{R}{\kappa} \qquad (15)$$

ote that constants c_1, c_2 and c_3 have the meaning of elastic wave speeds, while d_1, d_2 and $_3$ do not represent wave speed, but are diffusive coefficients and parameter θ may be nderstood as a manmade damping coefficient as in the previous section.

onsider an icosahedral quasicrystal specimen with a Griffith crack shown in Fig. 1, all arameters of geometry and loading are the same with those given in the previous, but in ie boundary conditions there are some different points, which are given as below

$$u_x = 0, \sigma_{yx} = 0, \sigma_{zx} = 0, w_x = 0, H_{yx} = 0, H_{zx} = 0 \qquad \text{on } x = 0 \text{ for } 0 \le y \le H$$

$$\sigma_{xx} = 0, \sigma_{yx} = 0, \sigma_{zx} = 0, H_{xx} = 0, H_{yx} = 0, H_{zx} = 0 \qquad \text{on } x = L \text{ for } 0 \le y \le H$$

$$\sigma_{yy} = p(t), \sigma_{xy} = 0, \sigma_{zy} = 0, H_{yy} = 0, H_{xy} = 0, H_{zy} = 0 \qquad \text{on } y = H \text{ for } 0 \le x \le L \qquad (16)$$

$$\sigma_{yy} = 0, \sigma_{xy} = 0, \sigma_{zy} = 0, H_{yy} = 0, H_{xy} = 0, H_{zy} = 0 \qquad \text{on } y = 0 \text{ for } 0 \le x \le a(t)$$

$$u_y = 0, \sigma_{xy} = 0, \sigma_{zy} = 0, w_y = 0, H_{xy} = 0, H_{zy} = 0 \qquad \text{on } y = 0 \text{ for } a(t) < x \le L$$

he initial conditions are

$$u_x(x,y,t)\big|_{t=0} = 0 \qquad u_y(x,y,t)\big|_{t=0} = 0 \qquad u_z(x,y,t)\big|_{t=0} = 0$$

$$w_x(x,y,t)\big|_{t=0} = 0 \qquad w_y(x,y,t)\big|_{t=0} = 0 \qquad w_z(x,y,t)\big|_{t=0} = 0 \qquad (17)$$

$$\frac{\partial u_x(x,y,t)}{\partial t}\bigg|_{t=0} = 0 \qquad \frac{\partial u_y(x,y,t)}{\partial t}\bigg|_{t=0} = 0 \qquad \frac{\partial u_z(x,y,t)}{\partial t}\bigg|_{t=0} = 0$$

.2 Some results

Ve now concentrate on investigating the phonon and phason fields in the icosahedral Al-d-Mn quasicrystal, in which we take $\rho = 5.1$ g/cm^3 and $\lambda = 74.2$ GPa, $\mu = 70.4$ GPa of the honon elastic moduli, for phason ones $K_1 = 72$ MPa, $K_2 = -37$ MPa and the onstant relevant to diffusion coefficient of phason is $_w = 1/\kappa = 4.8 \times 10^{-19}$ m$^3 \cdot$ s/kg=4.8×10^{-10} cm$^3 \cdot \mu$s/g . On the phonon-phason coupling onstant, there is no measured result for icosahedral quasicrystals so far, we take $R/\mu = 0.01$ for quasicrystals, and $R/\mu = 0$ for "decoupled quasicrystals" or crystals.

he problem is solved by the finite difference method, the principle, scheme and algorithm are lustrated as those in the previous section, and shall not be repeated here. The testing for the hysical model, scheme, algorithm and computer program are similar to those given in Section 2. he numerical results for dynamic initiation of crack growth problem, the phonon and hason displacements are shown in Fig. 5.

he dynamic stress intensity factor $K_1(t)$ is defined by

$$K_1(t) = \lim_{x \to a_0^+} \sqrt{\pi(x - a_0)} \sigma_{yy}(x,0,t)$$

nd the normalized dynamics stress intensity factor (D.S.I.F.) $\tilde{K}_1(t) = K_1(t)/\sqrt{\pi a_0} p_0$ is used, ie results are illustrated in Fig. 6, in which the comparison with those of crystals are shown, ne can see the effects of phason and phonon-phason coupling are evident very much.

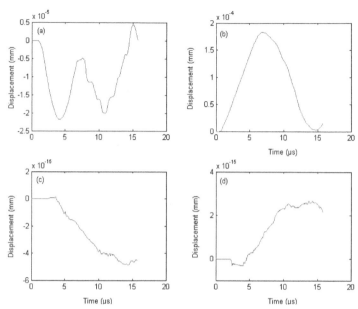

Fig. 5. Displacement components of quasicrystals versus time.
(a)displacement component u_x ; (b)displacement component u_y ;
(c)displacement component w_x ;(d)displacement component w_y

For the fast crack propagation problem the primary results are listed only the dynamic stress intensity factor versus time as below

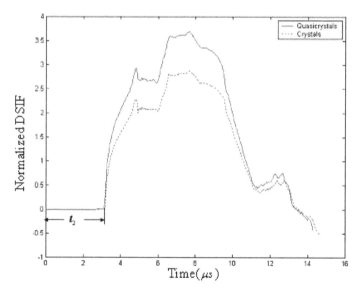

Fig. 6. Normalized dynamic stress intensity factor of central crack specimen under impact loading versus time

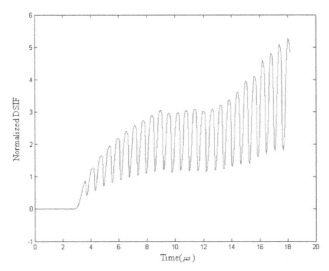

Fig. 7. Normalized stress intensity factor of propagating crack with constant crack speed versus time.

Details of this work can be given by Fan and co-workers [1], [10].

4. Conclusion and discussion

In Sections 1 through 3 a new model on dynamic response of quasicrystals based on argument of Lubensky et al is formulated. This model is regarded as an elasto-hydrodynamics model for the material, or as a collaborating model of wave propagation and diffusion. This model is more complex than pure wave propagation model for conventional crystals, the analytic solution is very difficult to obtain, except a few simple examples introduced in Fan's monograph [1]. Numerical procedure based on finite difference algorithm is developed. Computed results confirm the validity of wave propagation behaviour of phonon field, and behaviour of diffusion of phason field. The interaction between phonons and phasons are also recorded.

The finite difference formalism is applied to analyze dynamic initiation of crack growth and crack fast propagation for two-dimensional decagonal Al-Ni-Co and three-dimensional icosahedral Al-Pd-Mn quasicrystals, the displacement and stress fields around the tip of stationary and propagating cracks are revealed, the stress present singularity with order $r^{-1/2}$, in which r denotes the distance measured from the crack tip. For the fast crack propagation, which is a nonlinear problem — moving boundary problem, one must provide additional condition for determining solution. For this purpose we give a criterion for checking crack propagation/crack arrest based on the critical stress criterion. Application of this additional condition for determining solution has helped us to achieve the numerical simulation of the moving boundary value problem and revealed crack length-time evolution. However, more important and difficult problems are left open for further study.

Up to now the arguments on the physical meaning of phason variables based on hydrodynamics within different research groups have not been ended yet, see e.g. Coddens [11], which may be solved by further experimental and theoretical investigations.

5. References

[1] Fan T Y, 2010, Mathematical Theory of Elasticity of Quasicrystals and Its Applications, Beijing:Science Press/Heidelberg:Springer-Verlag.

[2] Lubensky T C , Ramaswamy S and Joner J, 1985, Hydrodynamics of icosahedral quasicrystals, Phys. Rev. B, 32(11), 7444.

[3] Socolar J E S, Lubensky T C and Steinhardt P J, 1986, Phonons, phasons and dislocations in quasicrystals, Phys. Rev. B, 34(5), 3345.

[4] Rochal S B and Lorman V L, 2002 , Minimal model of the phonon-phason dynamics on icosahedral quasicrystals and its application for the problem of internal friction in the i-AlPdMn alloys, Phys. Rev. B, 66 (14), 144204.

[5] Fan T Y , Wang X F, Li W et al., 2009, Elasto-hydrodynamics of quasicrystals, Phil. Mag., 89(6),501.

[6] Chernikov M A, Ott H R, Bianchi A et al., 1998, Elastic moduli of a single quasicrystal of decagonal Al-Ni-Co: evidence for transverse elastic isotropy, Phys. Rev. Lett. 80(2), 321-324.

[7] H. C. Jeong and P. J. Steinhardt, 1993, Finite-temperature elasticity phase transition in decagonal quasicrystals , Phys. Rev. B 48(13), 9394.

[8] Walz C, 2003, Zur Hydrodynamik in Quasikristallen, Diplomarbeit, Universitaet Stuttgart.

[9] Zhu A Y and Fan T Y, 2008, Dynamic crack propagation in a decagonal Al-Ni-Co quasicrystal , J. Phys.: Condens. Matter, 20(29), 295217.

[10] Wang X F, Fan T Y and Zhu A Y, 2009, Dynamic behaviour of the icosahedral Al-Pd-Mn quasicrystal with a Griffith crack, Chin Phys B, 18 (2), 709.(or referring to Zhu A Y: Study on analytic and numerical solutions in elasticity of three-dimensional quasicrystals and elastodynamics of two- and three-dimensional quasicrystals, Dissertation, Beijing Institute of Technology, 2009)

[11] Coddens G, 2006, On the problem of the relation between phason elasticity and phason dynamics in quasicrystals, Eur. Phys. J. B, 54(1), 37.

6

Hydrodynamics of a Droplet in Space

Department of Earth Planetary Materials Science,
Graduate School of Science,
Tohoku University
Japan

1. Introduction

1.1 Droplet in space

It is considered that our solar system 4.6 billion years ago was composed of a proto-sun and the circum-sun gas disk. In the gas disk, originally micron-sized fine dust particles accumulated by mutual collisions to be 1000 km-sized objects like as planets. Therefore, to understand the planet formation, we have to know the evolution of the dust particles in the early solar gas disk. One of the key materials is a millimeter-sized and spherical-shaped grain termed as "chondrule" observed in chondritic meteorites.

Chondrules are considered to have been formed from molten droplets about 4.6 billion years ago in the solar gas disk (Amelin et al., 2002; Amelin & Krot, 2007). Fig. 1 is a schematic of the formation process of chondrules. In the early solar gas disk, aggregation of the micron-sized dust particles took place before planet formation (Nakagawa et al., 1986). When the dust aggregates grew up to about 1 mm in size (precursor), some astrophysical process heated them to the melting point of about $1600 - 2100$ K (Hewins & Radomsky, 1990). The molten dust aggregate became a sphere by the surface tension (droplet), and then cooled again to solidify in a short period of time (chondrule). The formation conditions of chondrules, such as heating duration, maximum temperature, cooling rate, and so forth, have been investigated experimentally by many authors (Blander et al., 1976; Fredriksson & Ringwood, 1963; Harold C. Connolly & Hewins, 1995; Jones & Lofgren, 1993; Lofgren & Russell, 1986; Nagashima et al., 2006; Nelson et al., 1972; Radomsky & Hewins, 1990; Srivastava et al., 2010; Tsuchiyama & Nagahara, 1981-12; Tsuchiyama et al., 1980; 2004; Tsukamoto et al., 1999). However, it has been controversial what kind of astronomical event could have produced chondrules in early solar system. The chondrule formation is one of the most serious unsolved problems in planetary science.

The most plausible model for chondrule formation is a shock-wave heating model, which has been tested by many theoreticians (Ciesla & Hood, 2002; Ciesla et al., 2004; Desch & Jr., 2002; Hood, 1998; Hood & Horanyi, 1991; 1993; Iida et al., 2001; Miura & Nakamoto, 2006; Miura et al., 2002; Morris & Desch, 2010; Morris et al., 2009; Ruzmaikina & Ip, 1994; Wood, 1984). Fig. 2 is a schematic of dust heating mechanism by the shock-wave heating model. Initially, the chondrule precursors were floating in the gas disk without any large relative velocity against the ambient gas (panel (a)). When a shock wave was generated in the gas disk, the gas behind the shock front was accelerated suddenly. On the other hand, the chondrule

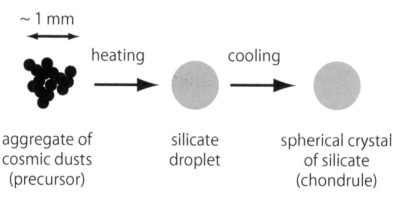

Fig. 1. Schematic of formation process of a chondrule. The precursor of chondrule is an aggregate of μm-sized cosmic dusts. The precursor is heated and melted by some mechanism, becomes a sphere by the surface tension, then cools to solidify in a short period of time.

precursors remain un-accelerated because of their inertia. Therefore, after passage of the shock front, the large relative velocity arises between the gas and dust particles (panel (b)). The relative velocity can be considered as fast as about 10 km s^{-1} (Iida et al., 2001). When the gas molecule collides to the surface of chondrule precursors with such large velocity, its kinetic energy thermalizes at the surface and heats the chondrule precursors, as termed as a gas drag heating. The peak temperature of the precursor is determined by the balance between the gas drag heating and the radiative cooling at the precursor surface (Iida et al., 2001). The gas drag heating is capable to heat the chondrule precursors up to the melting point if we consider a standard model of the early solar gas disk (Iida et al., 2001).

1.2 Physical properties of chondrules

The chondrule formation models, including the shock-wave heating model, are required not only to heat the chondrule precursors up to the melting point but also to reproduce other physical and chemical properties of chondrules recognized by observations and experiments. These properties that should be reproduced are summarized as observational constraints (Jones et al., 2000). The reference listed 14 constraints for chondrule formation. To date, there is no chondrule formation model that can account for all of these constraints.

Here, we review two physical properties of chondrules; size distribution and three-dimensional shape. The latter was not listed as the observational constraints in the literature (Jones et al., 2000), however, we would like to include it as an important constraint for chondrule formation. As discussed in this chapter, these two properties strongly relate to the hydrodynamics of molten chondrule precursors in the gas flow behind the shock front.

1.2.1 Size distribution

Fig. 3 shows the size distribution of chondrules compiled from measurement data in some literatures (Nelson & Rubin, 2002; Rubin, 1989; Rubin & Grossman, 1987; Rubin & Keil, 1984). The horizontal axis is the diameter D and the vertical axis is the cumulative fraction of

(a) before precursor meets shock front

chondrule precursor   shock front

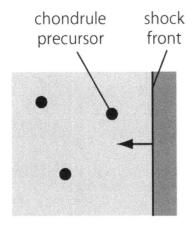

precursors in gas disk
w/o any relative velocity

(b) after precursor meets shock front

only gas is accelerated

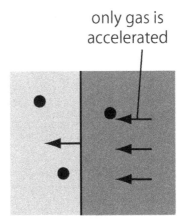

large relative velocity
between gas and precursor

Fig. 2. Schematic of the shock-wave heating model for chondrule formation. (a) The precursors of chondrules are in a gas disk around the proto-sun 4.6 billion years ago. The gas and precursors rotate around the proto-sun with almost the same angular velocity, so there is almost no relative velocity between the gas and precursors. (b) If a shock wave is generated in the gas disk by some mechanism, the gas behind the shock front is suddenly accelerated. In contrast, the precursor is not accelerated because of its large inertia. The difference of their behaviors against the shock front causes a large relative velocity between them. The precursors are heated by the gas friction in the high velocity gas flow.

chondrules smaller than D in diameter. Table 1 shows the mean diameter and the standard deviation of each measurement. It is found that the chondrule sizes vary according to chondrite type. The mean diameters of chondrules in ordinary chondrites (LL3 and L3) are from 600 μm to 1000 μm. In contrast, ones in enstatite chondrite (EH3) and carbonaceous chondrite (CO3) are from 100 μm to 200 μm.

It should be noted that the true chondrule diameters are slightly larger than the data shown in Fig. 3 and Table 1 because of the following reason. This data was obtained by observations on thin-sections of chondritic meteorites. The chondrule diameter on the thin-section is not necessarily the same as the true one because the thin-section does not always intersect the center of the chondrule. Statistically, the mean and median diameters measured on the thin section are, respectively, $\sqrt{2}/3$ and $\sqrt{3}/4$ of the true diameters (Hughes, 1978). However, we do not take care the difference between true and measured diameters because it is not a substantial issue in this chapter.

It is considered that in the early solar gas disk the dust aggregates have the size distribution from $\approx \mu$m (initial fine dust particles) to a few 1000 km (planets). In spite of the wide

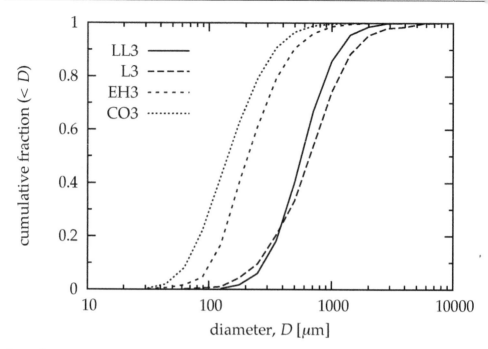

Fig. 3. Size distributions of natural chondrules in various types of chondritic meteorites (LL3, L3, EH3, and CO3). The vertical axis is the normalized cumulative number of chondrules whose diameters are smaller than that of the horizontal axis. Each data was compiled from the following literatures; LL3 chondrites (Nelson & Rubin, 2002), L3 chondrites (Rubin & Keil, 1984), EH3 chondrites (Rubin & Grossman, 1987), and CO3 chondrites (Rubin, 1989), respectively. The total number of chondrules measured in each literature is 719 for LL3, 607 for L3, 689 for EH3, and 2834 for CO3, respectively.

size range of solid materials, sizes of chondrules distribute in a very narrow range of about $100 - 1000$ μm. Two possibilities for the origin of chondrule size distribution can be considered; (i) size-sorting prior to chondrule formation, and (ii) size selection during chondrule formation. In the case of (i), we need some mechanism of size-sorting in the early solar gas disk (Teitler et al., 2010, and references therein). In the case of (ii), the chondrule formation model must account for the chondrule size distribution. The latter possibility is what we investigate in this chapter.

1.2.2 Deformation from a perfect sphere

It is considered that spherical chondrule shapes were due to the surface tension when they melted. However, their shapes deviate from a perfect sphere and the deviation is an important clue to identify the formation mechanism. Tsuchiyama et al. (Tsuchiyama et al., 2003) measured the three-dimensional shapes of chondrules using X-ray microtomography. They selected 20 chondrules with perfect shapes and smooth surfaces from 47 ones for further analysis. Their external shapes were approximated as three-axial ellipsoids with axial radii of a, b, and c ($a \geq b \geq c$), respectively. Fig. 4 shows results of the measurement. The horizontal

chondrite type	meteorite type	chondrule type*	number	diam. D [μm]	ref.
L3	Inman	BO	173	1038±937	(Rubin & Keil, 1984)
L3	Inman	RP+C	201	852±598	(Rubin & Keil, 1984)
L3	ALHA77011	BO	163	680±625	(Rubin & Keil, 1984)
L3	ALHA77011	RP+C	70	622±453	(Rubin & Keil, 1984)
LL3	total of 5 types	all	719	574^{+405}_{-237}	(Nelson & Rubin, 2002)
EH3	total of 3 types	all	689	219^{+189}_{-101}	(Rubin & Grossman, 1987)
CO3	total of 11 types	all	2834	148^{+132}_{-70}	(Rubin, 1989)

Table 1. Diameters of chondrules from various types of chondritic meteorites and the standard deviations. *BO = barred olivine, RP = radial pyroxene, C = cryptocrystalline. all = all types are included.

and vertical axes are axial ratios of b/a and c/b, respectively. A point $(b/a, c/b) = (1, 1)$ means a perfect sphere because all of three axes have the same length. As going downward from the point, the shape becomes oblate (disk-like shape) because $a = b > c$. On the other hand, the shape becomes prolate (rugby-ball-like shape) as going leftward because $a > b = c$. The chondrule shapes in the measurement are classified into two groups: spherical chondrules in group-A and prolate chondrules in group-B. Chondrules in group-A have axial ratios of $c/b >\sim 0.9$ and $b/a >\sim 0.9$. In contrast, chondrules in group-B have smaller values of b/a as $\approx 0.7 - 0.8$.

It is considered that the deviation from a perfect sphere results from the deformation of a molten chondrule before solidification. For example, if the molten chondrule rotates rapidly, the shape becomes oblate due to the centrifugal force (Chandrasekhar, 1965). However, the shapes of chondrules in group-B are prolate rather than oblate. Tsuchiyama et al. (Tsuchiyama et al., 2003) proposed that the prolate chondrules in group-B can be explained by spitted droplets due to the shape instability with high-speed rotation. However, it is not clear whether the transient process such as the shape instability accounts for the range of axial ratio of group-B chondrules or not.

4.3 Hydrodynamics of molten chondrule precursors

If chondrules were melted behind the shock front, the molten droplet ought to be exposed to the high-velocity gas flow. The gas flow causes many hydrodynamics phenomena on the molten chondrule droplet as follows. (i) Deformation: the ram pressure deforms the droplet shape from a sphere. (ii) Internal flow: the shearing stress at the droplet surface causes fluid flow inside the droplet. (iii) Fragmentation: a strong gas flow will break the droplet into many tiny fragments. Hydrodynamics of the droplet in high-velocity gas flow strongly relates to the physical properties of chondrules. However, these hydrodynamics behaviors have not been investigated in the framework of the chondrule formation except of a few examples that neglected non-linear effects of hydrodynamics (Kato et al., 2006; Sekiya et al., 2003; Uesugi et al., 2005; 2003).

To investigate the hydrodynamics of a molten chondrule droplet in the high-velocity gas flow, we performed computational fluid dynamics (CFD) simulations based on cubic-interpolated propagation/constrained interpolation profile (CIP) method. The CIP method is one of the high-accurate numerical methods for solving the advection equation (Yabe & Aoki, 1991;

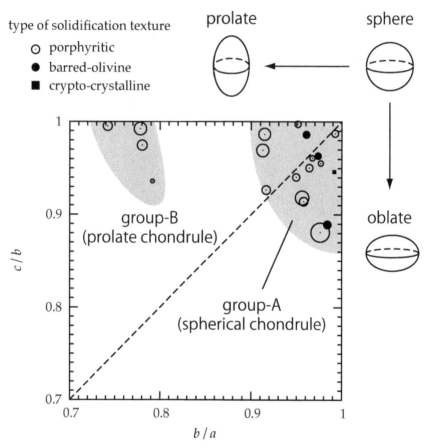

Fig. 4. Three-dimensional shapes of chondrules (Tsuchiyama et al., 2003, and their unpublished data). a, b, and c are axial radii of chondrules when their shapes are approximated as three-axial ellipsoids ($a \geq b \geq c$), respectively. The textures of these chondrules are 16 porphyritic (open circle), 3 barred-olivine (filed circle), and 1 crypto-crystalline (filled square). The radius of each symbol is proportional to the effective radius of each chondrule $r_* \equiv (abc)^{1/3}$; the largest circle corresponds to $r_* = 1129$ ˉm. For the data of crypto-crystalline, $r_* = 231$ ˉm. Chondrule shapes are classified into two groups: group-A shows the relatively small deformation from the perfect sphere, and group-B is prolate with axial ratio of $b/a \approx 0.7 - 0.8$.

'abe et al., 2001). It can treat both compressible and incompressible fluids with large density atios simultaneously in one program (Yabe & Wang, 1991). The latter advantage is important or our purpose because the droplet density ($\approx$ 3 g cm^{-3}) differs from that of the gas disk $\approx 10^{-8}$ g cm^{-3} or smaller) by many orders of magnitude.

n addition, we should pay a special attention how to model the ram pressure of the gas flow. "he gas around the droplet is so rarefied that the mean free path of the gas molecules is an •rder of about 100 cm if we consider a standard gas disk model. The mean free path is much arger than the typical size of chondrules. This means that the gas flow around the droplet is ι free molecular flow, so it does not follow the hydrodynamical equations. Therefore, in our nodel, the ram pressure acting on the droplet surface per unit area is explicitly given in the •quation of motion for the droplet by adopting the momentum flux method as described in ection 3.2.2.

I.4 Aim of this chapter

The hydrodynamical behaviors of molten chondrules in a high-velocity gas flow are important o elucidate the origin of physical properties of chondrules. However, it is difficult for ›xperimental studies to simulate the high-velocity gas flow in the early solar gas disk, vhere the gas density is so rarefied that the gas flow around droplets does not follow the ιydrodynamics equations. We developed the numerical code to simulate the droplet in a ιigh-velocity rarefied gas flow. In this chapter, we describe the details of our hydrodynamics :ode and the results. We propose new possibilities for the origins of size distribution and three-dimensional shapes of chondrules based on the hydrodynamics simulations.

Ne describe the governing equations in section 2 and the numerical procedures in section ›. In section 4, we describe the results of the hydrodynamics simulations regarding the leformation of molten chondrules in the high-velocity rarefied gas flow and discuss the ›rigin of rugby-ball-like shaped chondrules. In section 5, we describe the results regarding he fragmentation of molten chondrules and consider the relation to the size distribution of :hondrules. We conclude our hydrodynamics simulations in section 6.

2. Governing equations

The governing equations are the equation of continuity and the Navier-Stokes equation as ›ollows;

$$\frac{\partial \rho}{\partial t} + \vec{\nabla} \cdot (\rho \vec{u}) = 0, \tag{1}$$

$$\frac{\partial \vec{u}}{\partial t} + (\vec{u} \cdot \vec{\nabla})\vec{u} = \frac{-\vec{\nabla} p + \mu \nabla^2 \vec{u} + \vec{F}_s + \vec{F}_g}{\rho} + \vec{g}, \tag{2}$$

where ρ is the density of fluid, $\vec{u}$ is the velocity, p is the pressure, and μ is the viscosity. The ·am pressure of the high-velocity gas flow, $\vec{F}_g$, is exerted on the surface of the droplet and ;iven by (Sekiya et al., 2003)

$$\vec{F}_g = -p_{fm}(\vec{n}_i \cdot \vec{n}_g)\vec{n}_g\delta(\vec{r} - \vec{r}_i) \quad \text{for } \vec{n}_i \cdot \vec{n}_g \leq 0, \tag{3}$$

where $\vec{n}_i$ is the unit normal vector of the surface of the droplet, $\vec{n}_g$ is the unit vector pointing :he direction in which the gas flows, and $\vec{r}_i$ is the position of the liquid-gas interface. The lelta function $\delta(\vec{r} - \vec{r}_i)$ means that the ram pressure works only at the interface. The ram pressure does not work for $\vec{n}_i \cdot \vec{n}_g > 0$ because it indicates the opposite surface which does not face the molecular flow. The ram pressure causes the deceleration of the center of mass of the droplet. In our coordinate system co-moving with the center of mass, the apparent gravitational acceleration $\vec{g}$ should appear in the equation of motion. The surface tension, $\vec{F}_s$, is given by (Brackbill et al., 1992)

$$\vec{F}_s = -\gamma_s \kappa \vec{n}_i \delta(\vec{r} - \vec{r}_i), \tag{4}$$

where γ_s is the fluid surface tension and κ is the local surface curvature. Finally, we consider the equation of state given by

$$\frac{dp}{d\rho} = c_s^2, \tag{5}$$

where c_s is the sound speed.

3. Numerical methods in hydrodynamics

To solve the equation of continuity (Eq. (1)) numerically, we introduce a color function ϕ that changes from 0 to 1 continuously. For incompressible two fluids, a density in each fluid is uniform and has a sharp discontinuity at the interface between these two fluids if the density of a fluid is different from another one. By using the color function, we can distinguish these two fluids as follows; $\phi = 1$ for fluid 1, $\phi = 0$ for fluid 2, and a region where $0 < \phi < 1$ for the interface. The density of a fluid element is given by

$$\rho = \phi \rho_1 + (1 - \phi)\rho_2, \tag{6}$$

where ρ_1 and ρ_2 are the inherent densities for fluid 1 and fluid 2, respectively. The governing equation for ϕ is given by

$$\frac{\partial \phi}{\partial t} + \vec{\nabla} \cdot (\phi \vec{u}) = 0. \tag{7}$$

The conservation equation for ϕ (Eq. (7)) is approximately equivalent to the original one (Eq. (1)) through the relationship between ρ and ϕ given by Eq. (6) (Miura & Nakamoto, 2007). Therefore, the problem to solve Eq. (1) results in to solve Eq. (7). We solve Eq. (7) using R-CIP-CSL2 method with anti-diffusion technique (sections 3.1.2 and 3.1.3).

In this study, the fluid 1 is the molten chondrule and the fluid 2 is the disk gas around the chondrule. The inherent densities are given by $\rho_1 = \rho_d$ and $\rho_2 = \rho_a$, where subscripts "d" and "a" mean the droplet and ambient gas, respectively. The other physical values of the fluid element (viscosity μ and sound speed c_s) are given in the same manner as the density ρ, namely, $\mu = \phi \mu_d + (1 - \phi)\mu_a$ and $c_s = \phi c_{s,d} + (1 - \phi)c_{s,a}$, respectively.

The Navier-Stokes equation (Eq. (2)) and the equation of state (Eq. (5)) are separated into two phases; the advection phase and the non-advection phase. The advection phases are written as

$$\frac{\partial \vec{u}}{\partial t} + (\vec{u} \cdot \vec{\nabla})\vec{u} = 0,$$

$$\frac{\partial p}{\partial t} + (\vec{u} \cdot \vec{\nabla})p = 0. \tag{8}$$

Parameter	Sign	Value
Momentum of gas flow	p_{fm}	4000 dyn cm^{-2}
Surface tension	γ_s	400 dyn cm^{-1}
Viscosity of droplet	μ_d	1.3 g cm^{-1} s^{-1}
Density of droplet	ρ_d	3 g cm^{-3}
Sound speed of droplet	$c_{s,d}$	2×10^5 cm s^{-1}
Density of ambient	ρ_a	10^{-6} g cm^{-3}
Sound speed of ambient	$c_{s,a}$	10^{-5} cm s^{-1}
Viscosity of ambient	μ_a	10^{-2} g cm^{-1} s^{-1}
Droplet radius	r_0	500 μm

Table 2. Canonical input physical parameters for simulations of molten chondrules exposed to the high-velocity rarefied gas flow. We ought to use these parameters if there is no special description.

We solve above equations using the R-CIP method, which is the oscillation preventing method for advection equation (section 3.1.1). The non-advection phases can be written as

$$\frac{\partial \vec{u}}{\partial t} = -\frac{\vec{\nabla} p}{\rho} + \frac{\vec{Q}}{\rho},$$

$$\frac{\partial p}{\partial t} = -\rho c_s^2 \vec{\nabla} \cdot \vec{u}, \tag{9}$$

where $\vec{Q}$ is the summation of forces except for the pressure gradient. The problem intrinsic in incompressible fluid is in the high sound speed in the pressure equation. Yabe and Wang (Yabe & Wang, 1991) introduced an excellent approach to avoid the problem (section 3.2.1). It is called as the C-CUP method (Yabe & Wang, 1991). The numerical methods to calculate ram pressure of the gas flow and the surface tension of droplet in $\vec{Q}$ are described in sections 3.2.2 and 3.2.3, respectively.

The input parameters adopted in this chapter are listed in Table 2.

3.1 Advection phase

3.1.1 CIP method

The CIP method is one of the high-accurate numerical methods for solving the advection equation (Yabe & Aoki, 1991; Yabe et al., 2001). In one-dimension, the advection equation is written as

$$\frac{\partial f}{\partial t} + u \frac{\partial f}{\partial x} = 0, \tag{10}$$

where f is a scaler variable of the fluid (e.g., density), u is the fluid velocity in the x-direction, and t is the time. When the velocity u is constant, the exact solution of Eq. (10) is given by

$$f(x;t) = f(x - ut;0), \quad \text{when } u \text{ is constant}, \tag{11}$$

which indicates a simple translational motion of the spatial profile of f with the constant velocity u.

Let us consider that the values of f on the computational grid points x_{i-1}, x_i, and x_{i+1} are given at the time step n and denoted by f_{i-1}^n, f_i^n, and f_{i+1}^n, respectively. In Fig. 5, f^n are shown

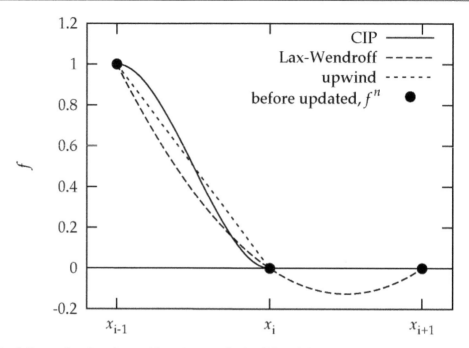

Fig. 5. Interpolate functions with various methods: CIP (solid), Lax-Wendroff (dashed), and first-order upwind (dotted). The filled circles indicate the values of f defined on the digitized grid points x_{i-1}, x_i, and x_{i+1} before updated.

by filled circles. From Eq. (11), we can obtain the values of f_i at the next time step $n + 1$ by just obtaining f_i^n at the upstream point $x = x_i - u\Delta t$, where Δt is the time interval between t^n and t^{n+1}. If the upstream point is not exactly on the grid points, which is a very usual case, we have to interpolate f_i^n with an appropriate mathematical function composed of f_{i-1}^n, f_i^n, and so forth. There are some variations of the numerical solvers by the difference of the interpolate function $F_i(x)$. One of them is the first-order upwind method, which interpolates f_i^n by a linear function and satisfies following two constraints; $F_i(x_{i-1}) = f_{i-1}^n$ and $F_i(x_i) = f_i^n$ (here we assume that $u > 0$ and the upstream point for f_i^n locates left-side of x_i). The other variation is the Lax-Wendroff method, which uses a quadratic polynomial satisfying three constraints; $F_i(x_{i-1}) = f_{i-1}^n$, $F_i(x_i) = f_i^n$, and $F_i(x_{i+1}) = f_{i+1}^n$. We show these interpolation functions in Fig. 5.

On the contrary, the CIP method interpolates using a cubic polynomial, which satisfies following four constraints; $F_i(x_{i-1}) = f_{i-1}^n$, $F_i(x_i) = f_i^n$, $\partial F_i / \partial x(x_{i-1}) = f_{x,i-1}^n$, and $\partial F_i / \partial x(x_i) = f_{x,i}^n$, where $f_x \equiv \partial f / \partial x$ is the spatial gradient of f. The interpolation function is given by

$$F_i(x) = a_i(x - x_i)^3 + b_i(x - x_i)^2 + c_i(x - x_i) + d_i, \tag{12}$$

where a_i, b_i, c_i, and d_i are the coefficients determined from f_{i-1}^n, f_i^n, $f_{x,i-1}^n$, and $f_{x,i}^n$. The expressions of these coefficients are shown in (Yabe & Aoki, 1991). We show the profile of $F_i(x)$ in Fig. 5 with $f_{x,i-1}^n = f_{x,i}^n = 0$. In the CIP method, therefore, we need the values of f_x^n in addition of f^n for solving the advection phase.

In the CIP method, f_x is treated as an independent variable and updated independently from f as follows. Differentiating Eq. (10) with respect to x, we obtain

$$\frac{\partial f_x}{\partial t} + u\frac{\partial f_x}{\partial x} = -f_x\frac{\partial u}{\partial x}, \tag{13}$$

where the second term of the left-hand side indicates the advection term and the right-hand side indicates the non-advection term. The interpolate function for the advection of f_x is given by $\partial F_i/\partial x$. The non-advection term can be solved analytically by considering that $\partial u/\partial x$ is constant.

Additionally, there is an oscillation preventing method in the concept of the CIP method, in which the rational function is used as the interpolate function. The rational function is written as (Xiao et al., 1996)

$$F_i(x) = \frac{a_i(x - x_i)^3 + b_i(x - x_i)^2 + c_i(x - x_i) + d_i}{1 + \alpha_i\beta_i(x - x_i)}, \tag{14}$$

where α_i and β_i are coefficients. The expressions of these coefficients are shown in (Xiao et al., 1996). Usually, we adopt $\alpha_i = 1$ to prevent oscillation. This method is called as the R-CIP method. The model with $\alpha_i = 0$ corresponds to the normal CIP method.

3.1.2 CIP-CSL2 method

The CIP-CSL2 method is one of the numerical methods for solving the conservative equation. In one-dimension, the conservative equation is written as

$$\frac{\partial f}{\partial t} + \frac{\partial(uf)}{\partial x} = 0. \tag{15}$$

Integrating Eq. (15) over x from x_i to x_{i+1}, we obtain

$$\frac{\sigma_{i+1/2}}{\partial t} + [uf]_{x_i}^{x_{i+1}} = 0, \tag{16}$$

where $\sigma_{i+1/2} \equiv \int_{x_i}^{x_{i+1}} f dx$. For f being density, $\sigma_{i+1/2}$ corresponds to the mass contained in a computational cell between i and $i + 1$, so it should be conserved during the time integration. Since the physical meaning of uf in the second term of the left-hand side is the flux of σ per unit area and per unit time, the time evolution of σ is determined by

$$\sigma_{i+1/2}^{n+1} = \sigma_{i+1/2}^n - J_{i+1} + J_i, \tag{17}$$

where $J_i \equiv \int_{t^n}^{t^{n+1}} uf dt$ is the transported value of σ from a region of $x < x_i$ to that of $x > x_i$ within Δt. The CIP-CSL2 method uses the integrated function $D_i(x) \equiv \int_{x_{i-1}}^x F_i(x) dx$ for the interpolation when $u_i > 0$. The function $D_i(x)$ is a cubic polynomial satisfying following four constraints; $D_i(x_{i-1}) = 0$, $D_i(x_i) = \sigma_{i-1/2}$, $\partial D_i/\partial x(x_{i-1}) = F_i(x_{i-1}) = f_{i-1}$, and $\partial D_i/\partial x(x_i) = F_i(x_i) = f_i$. Moreover, since Eq. (15) can be rewritten as the same form of Eq. (13), we can obtain the updated value f^{n+1} as well as f_x^{n+1} in the CIP method.

Additionally, there is an oscillation preventing method in the concept of the CIP-CSL2 method, in which the rational function is used for the function $D_i(x)$ (Nakamura et al., 2001). This method is called as the R-CIP-CSL2 method.

3.1.3 Anti-diffusion

To keep the sharp discontinuity in the profile of ϕ, we explicitly add an diffusion term with a negative diffusion coefficient α (anti-diffusion) to the CIP-CSL2 method (Miura & Nakamoto, 2007). In our model, we have an additional diffusion equation about σ as

$$\frac{\partial \sigma}{\partial t} = \frac{\partial}{\partial x}\left(\alpha \frac{\partial \sigma}{\partial x}\right).$$ (18)

Eq. (18) can be separated into two equations as

$$\frac{\partial \sigma}{\partial t} = -\frac{\partial J'}{\partial x},$$ (19)

$$J' = -\alpha \frac{\partial \sigma}{\partial x},$$ (20)

where J' indicates the anti-diffusion flux per unit area and per unit time. Using the finite difference method, we obtain

$$\sigma^{**}_{i+1/2} = \sigma^{*}_{i+1/2} - (\hat{J}'_{i+1} - \hat{J}'_{i}),$$ (21)

$$\hat{J}'_{i} = -\hat{\alpha}_{i} \times \mathrm{minmod}(S_{i-1}, S_{i}, S_{i+1}),$$ (22)

where $\hat{J} \equiv J'/(\Delta x/\Delta t)$ is the mass flux which has the same dimension of σ, $\hat{\alpha} \equiv \alpha/(\Delta x^2/\Delta t)$ is the dimensionless diffusion coefficient, and $S_i \equiv \sigma_{i+1/2} - \sigma_{i-1/2}$. The superscripts * and ** indicate the time step just before and after the anti-diffusion. The minimum modulus function (minmod) is often used in the concept of the flux limiter and has a non-zero value of $\mathrm{sign}(a)\min(|a|, |b|, |c|)$ only when a, b, and c have the same sign. The value of the diffusion coefficient $\hat{\alpha}$ is also important. Basically, we take $\hat{\alpha} = -0.1$ for the anti-diffusion. Here, it should be noted that σ takes the limited value as $0 \leq \sigma \leq \sigma_{\mathrm{m}}$, where σ_{m} is the initial value for inside of the droplet. The undershoot ($\sigma < 0$) or overshoot ($\sigma > \sigma_{\mathrm{m}}$) are physically incorrect solutions. To avoid that, we replace $\hat{\alpha}_i = 0.1$ only when $\sigma_{i-1/2}$ or $\sigma_{i+1/2}$ are out of the appropriate range. We insert the anti-diffusion calculation after the CIP-CSL2 method is completed.

3.1.4 Test calculation

In order to demonstrate the advantage of the CIP method, we carried out one-dimensional advection calculations with various numerical methods. Fig. 6 shows the spatial profiles of f of the test calculations. The horizontal axis is the spatial coordinate x. The initial profile is given by the solid line, which indicates a rectangle wave. We set the fluid velocity $u = 1$, the intervals of the grid points $\Delta x = 1$, and the time step for the calculation $\Delta t = 0.2$. These conditions give the CFL number $\nu \equiv u\Delta t/\Delta x = 0.2$, which indicates that the profile of f moves 0.2 times the grid interval per time step. Therefore, the right side of the rectangle wave will reach $x = 80$ after 300 time steps and the dashed line indicates the exact solution. The filled circles indicate the numerical results after 300 time steps.

The upwind method does not keep the rectangle shape after 300 time steps and the profile becomes smooth by the numerical diffusion (panel a). In the Lax-Wendroff method, the numerical oscillation appears behind the real wave (panel b). Comparing with above two methods, the CIP method seems to show better solution, however, some undershoots ($f < 0$)

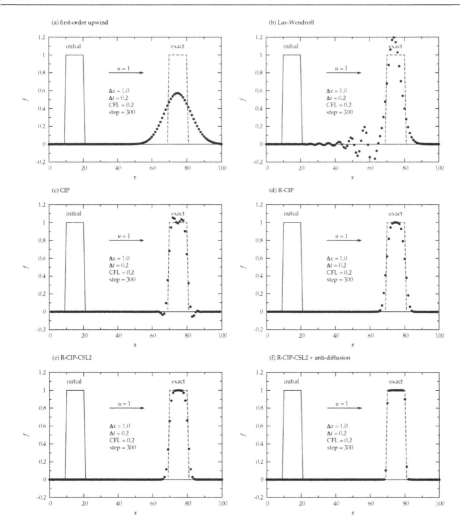

Fig. 6. Numerical solutions of the one-dimensional advection or conservation equation olved by various methods: (a) first-order upwind, (b) Lax-Wendroff, (c) CIP, (d) R-CIP, (e) R-CIP-CSL2 without anti-diffusion, and (f) R-CIP-CSL2 with anti-diffusion.

r overshoots ($f > 1$) are observed in the numerical result (panel c). In the R-CIP method, lthough the faint numerical diffusion has still remained, we obtain the excellent solution omparing with the above methods.

Ve also show the numerical results of the one-dimensional conservative equation. We use the ame conditions with the one-dimensional advection equation. Note that Eq. (15) corresponds Eq. (10) when the velocity u is constant. The panel (e) shows the result of the R-CIP-CSL2 nethod, which is similar to that of the R-CIP method. In the panel (f), we found that the ombination of the R-CIP-CSL2 method and the anti-diffusion technique shows the excellent olution in which the numerical diffusion is prevented effectively.

3.2 Non-advection phase

3.2.1 C-CUP method

Using the finite difference method to Eq. (9), we obtain (Yabe & Wang, 1991)

$$\frac{\vec{u}^{**} - \vec{u}^*}{\Delta t} = -\frac{\vec{\nabla} p^{**}}{\rho^*} + \frac{\vec{Q}}{\rho^*}, \quad \frac{p^{**} - p^*}{\Delta t} = -\rho^* c_s^2 \vec{\nabla} \cdot \vec{u}^{**}, \tag{23}$$

where the superscripts * and ** indicate the times before and after calculating the non-advection phase, respectively. Since the sound speed is very large in the incompressible fluid, the term related to the pressure should be solved implicitly. In order to obtain the implicit equation for p^{**}, we take the divergence of the left equation and substitute $\vec{u}^{**}$ into the right equation. Then we obtain an equation

$$\vec{\nabla} \cdot \left(\frac{\vec{\nabla} p^{**}}{\rho^*} \right) = \frac{p^{**} - p^*}{\rho^* c_s^2 \Delta t^2} + \frac{\vec{\nabla} \cdot \vec{u}^*}{\Delta t} + \vec{\nabla} \cdot \left(\frac{\vec{Q}}{\rho^*} \right). \tag{24}$$

The problem to solve Eq. (24) resolves itself into to solve a set of linear algebraic equations in which the coefficients becomes an asymmetric sparse matrix. After p^{**} is solved, we can calculate $\vec{u}^{**}$ by solving the left equation in Eq. (23).

3.2.2 Ram pressure of free molecular flow

The ram pressure of the gas flow is acting on the droplet surface exposed to the high-velocity gas flow. It should be noted that the gas flow around a mm-sized droplet does not follow the hydrodynamical equations because the nebula gas is too rarefied. The mean free path of the nebula gas can be estimated by $l = 1/(ns)$, where s is the collisional cross section of gas molecules and n is the number density of the nebula gas. Typically, we adopt $n \approx 10^{14}$ cm^{-3} based on the standard model of the early solar system at a distance from the sun of an astronomical unit (Hayashi et al., 1985). Substituting $s \approx 10^{-16}$ cm^{-2} for the hydrogen molecule (Hollenbach & McKee, 1979), we obtain $l \approx 100$ cm. On the other hand, the typical size of chondrules is about a few 100 μm (see Fig. 3). Since the object that disturbs the gas flow is much smaller than the mean free path of the gas, the free stream velocity field is not disturbed except of the direct collision with the droplet (free molecular flow).

Consider that the molecular gas flows for the positive direction of the x-axis. The x-component of the ram pressure $F_{g,x}$ is given by

$$F_{g,x} = p_{fm} \delta(x - x_i), \tag{25}$$

where x_i is the position of the droplet surface. This equation can be separated into two equations as

$$F_{g,x} = -\frac{\partial M}{\partial x}, \quad \frac{\partial M}{\partial x} = -p_{fm} \delta(x - x_i), \tag{26}$$

where M is the momentum flux of the molecular gas flow. The right equation in Eq. (26) means that the momentum flux terminates at the droplet surface. The left equation in Eq. (26) means that the decrease of the momentum flux per unit length corresponding to the ram pressure per unit area.

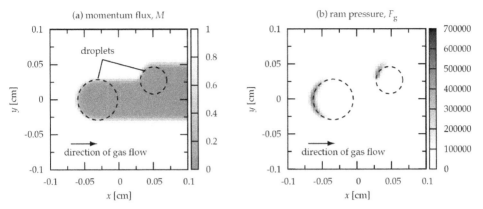

Fig. 7. Spatial distributions of the momentum flux M (a) and the ram pressure F_g (b) of the free molecular gas flow around a spherical droplet in xy-plane. The dashed circles are sections of the droplet surfaces in xy-plane. Units of the gray scales are p_{fm} for the panel (a) and dyn cm^{-3} for the panel (b), respectively. We adopt $p_{fm} = 5000$ dyn cm^{-2} in this figure.

Using the finite difference method to the right equation in Eq. (26), we obtain

$$M_{i+1} = M_i - p_{fm}(\bar{\phi}_{i+1} - \bar{\phi}_i) \quad \text{for } \bar{\phi}_{i+1} \geq \bar{\phi}_i, \tag{27}$$

where $\bar{\phi}$ is the smoothed profile of ϕ (see section 3.2.4), and $M_{i+1} = M_i$ for $\bar{\phi}_{i+1} < \bar{\phi}_i$ because the momentum flux does not increase when the molecular flow goes outward from inside of the droplet. Similarly, we obtain

$$F_{g,x_i} = -\frac{M_i - M_{i+1}}{\Delta x}, \tag{28}$$

from the left equation in Eq. (26). The momentum flux at upstream is $M_0 = p_{fm}$. First, we solve Eq. (27) and obtain the spatial distribution of the molecular gas flow in all computational domain. Then, we calculate the ram pressure by Eq. (28).

Fig. 7(a) shows the distribution of momentum flux M around two droplets in xy-plane. The dashed circles are the external shapes of large and small droplets. The gray scale is normalized by p_{fm}, so unity (white region) means undisturbed molecular flow and zero (dark region) means no flux because the free molecular flow is obstructed by the droplet. It is found that the gas flow is obstructed only behind the droplets. Fig. 7(b) shows the distribution of the ram pressure $F_{g,x}$ calculated from the momentum flux distribution. The ram pressure is acting at the droplet surface where M changes steeply. Note that no ram pressure acts at bottom half of the smaller droplet because the molecular flow is obstructed by the larger one. As shown in Fig. 7, the model of ram pressure shown here well reproduces the property of free molecular flow.

We calculate the momentum flux M and the ram pressure F_g at every time step in numerical simulations. Therefore, these spatial distributions are affected by droplet deformation.

3.2.3 Surface tension

The surface tension is the normal force per unit interfacial area. Brackbill et al. (Brackbill et al., 1992) introduced a method to treat the surface tension as a volume force by replacing the discontinuous interface to the transition region which has some width. According to them, the surface tension is expressed as

$$\vec{F}_s = \gamma_s \kappa \vec{\nabla} \phi / [\phi], \tag{29}$$

where $[\phi]$ is the jump in color function at the interface between the droplet and the ambient gas. In our definition, we obtain $[\phi] = 1$. The curvature is given by

$$\kappa = -(\vec{\nabla} \cdot \vec{n}), \tag{30}$$

where

$$\vec{n} = \vec{\nabla} \phi / |\vec{\nabla} \phi|. \tag{31}$$

The finite difference method of Eq. (31) is shown in (Brackbill et al., 1992). When we calculate the surface tension, we use the smoothed profile of ϕ (see section 3.2.4).

3.2.4 Smoothing

We can obtain the numerical results keeping the sharp interface between the droplet and the ambient region. However, the smooth interface is suitable for calculating the smooth surface tension. We use the smoothed profile of ϕ only at the time to calculate the surface tension and the ram pressure acting on the droplet surface. In this study, the smoothed color function $\bar{\phi}$ is calculated by

$$\bar{\phi} = \frac{1}{2}\phi_{i,j,k} + \frac{1}{2}\frac{\phi_{i,j,k} + C_1 \sum_{L_1}^{6} \phi_{L_1} + C_2 \sum_{L_2}^{12} \phi_{L_2} + C_3 \sum_{L_3}^{8} \phi_{L_3}}{1 + 6C_1 + 12C_2 + 8C_3}, \tag{32}$$

where L_1, L_2, and L_3 indicate grid indexes of the nearest, second nearest, and third nearest from the grid point (i, j, k), for example, $L_1 = (i+1, j, k)$, $L_2 = (i+1, j+1, k)$, $L_3 = (i+1, j+1, k+1)$, and so forth. It is easily found that in the three-dimensional Cartesian coordinate system, there are six for L_1, twelve for L_2, and eight for L_3, respectively. The coefficients are set as

$$C_1 = 1/(6 + 12/\sqrt{2} + 8/\sqrt{3}), \quad C_2 = C_1/\sqrt{2}, \quad C_3 = C_1/\sqrt{3}. \tag{33}$$

We iterate the smoothing five times. Then, we obtain the smooth transition region of about twice grid interval width. We use the smooth profile of ϕ only when calculating the surface tension and the ram pressure. It should be noted that the original profile ϕ with the sharp interface is kept unchanged.

4. Deformation of droplet by gas flow

4.1 Vibrational motion

We assume that the gas flow suddenly affects the initially spherical droplet. Fig. 8 shows the time sequence of the droplet shape and the internal velocity. The horizontal and vertical axes are the x- and y-axes, respectively. The solid line is the section of the droplet surface in xy-plane. Arrows show the velocity field inside the droplet. The gas flow comes from the left side of the panel. The panel (a) shows the initial condition for the calculation. The panel (b) shows a snapshot at $t = 0.55$ msec. The droplet begins to be deformed due to the gas ram pressure. The fluid elements at the surface layer, which is directly facing the gas flow, are blown to the downstream. In contrast, the velocity at the center of the droplet turns to

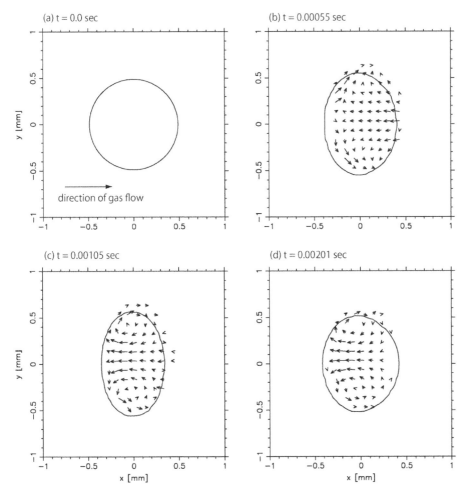

Fig. 8. Time evolution of molten droplet exposed to the gas flow. The gas flow comes from the left side of panels. We use $p_{fm} = 10^4$ dyn cm^{-2}, $r_0 = 500$ $^-$m, and $\mu_d = 1.3$ poise for calculations.

upstream of the gas flow because the apparent gravitational acceleration takes place in our coordinate system. The droplet continues to be deformed further, and after $t = 1.0$ msec, the degree of deformation becomes maximum (see panel (c)). After that, the droplet begins to recover its shape to the sphere due to the surface tension. The recovery motion is not all but almost over at the panel (d). The droplet repeats the deformation by the ram pressure and the recovery motion by the surface tension until the viscosity dissipates the internal motion of the droplet.

Fig. 9 shows the time variation of axial ratio c/b of the droplet. Each curve shows the calculation result for the different value of the ram pressure p_{fm}. The droplet is compressed unidirectionally by the gas flow, so the length of minor axis c corresponds to the half length of droplet axis in the direction of the gas flow. The axial ratio c/b is unity at the

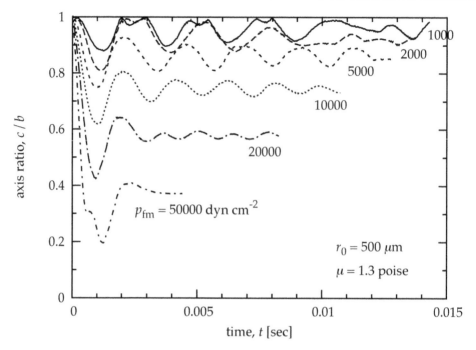

Fig. 9. Vibrational motions of molten droplet; the deformation by the ram pressure and the recovery motion by the surface tension. The horizontal axis is the time since the ram pressure begins to affect the droplet and the vertical axis is the axial ratio of the droplet c/b. Each curve shows the calculation result for the different value of the ram pressure p_{fm}. We use $r_0 = 500$ ¯m and $\mu_d = 1.3$ poise for calculations.

beginning because the initial droplet shape is a perfect sphere. The axial ratio decreases as time goes by because of the compression. After about 1 msec, c/b reaches minimum and then increases due to the surface tension. After this, the axial ratio vibrates with a constant frequency and finally the vibrational motion damps due to viscous dissipation. The calculated frequency of the vibrational motion is about 2 msec not depending on p_{fm}. The calculated frequency is consistent with that of a capillary oscillations of a spherical droplet given by $P_{vib} = 2\pi\sqrt{\rho_d r_0^3/8\gamma_s} \approx 2.15$ msec (Landau & Lifshitz, 1987).

4.2 Overdamping

Fig. 10 shows the time variation of the axial ratio c/b when the viscosity is 100 times larger than that in Fig. 9. It is found that the axial ratio converges on the value at steady state without any vibrational motion. This is an overdamping due to the strong viscous dissipation.

4.3 Effect of droplet rotation

We carried out the hydrodynamics simulations of non-rotating molten droplet in previous sections. However, the rotation of the droplet should be taken into consideration as the following reason. A chondrule before melting is an aggregate of numerous fine particles,

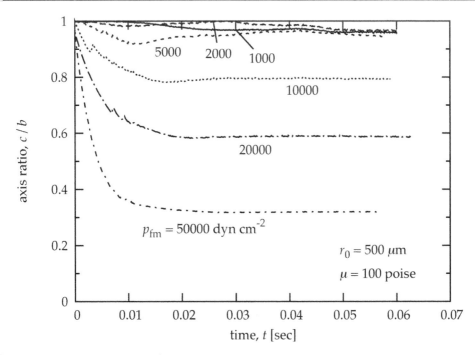

Fig. 10. Same as Fig. 9 except of $\mu_d = 100$ poise.

so the shape is irregular in general. The irregular shape causes a net torque in an uniform gas flow. Therefore, it is naturally expected that the molten chondrule also rotates at a certain angular velocity.

The angular velocity ω_f can be roughly estimated by $I\omega_f \approx N\Delta t$, where I is the moment of inertia of chondrule and Δt is the duration to receive the net torque N. Assuming that the small fraction f of the cross-section of the precursor contributes to produce the net torque N, we obtain $N \approx f\pi r_0^3 p_{fm}$. We can set $\Delta t \approx \pi/\omega_f$ (a half-rotation period) because the sign of N would change after half-rotation. Substituting $I = (8/15)\pi r_0^5 \rho_d$, which is the moment of inertia for a sphere with an uniform density ρ_d, we obtain the angular velocity (Miura, Nakamoto & Doi, 2008)

$$\omega_f \approx \sqrt{15 f \pi p_{fm}/8r_0^2\rho_d}$$

$$= 140 \left(\frac{f}{0.01}\right)^{1/2} \left(\frac{p_{fm}}{10^4 \text{ dyn cm}^{-2}}\right)^{1/2} \left(\frac{r_0}{1 \text{ mm}}\right)^{-1} \text{ rad s}^{-1}. \tag{34}$$

Therefore, in the shock-wave heating model, the droplet should be rotating rapidly if most of the angular momentum is maintained during melting.

In addition, it should be noted that the rotation axis is likely to be perpendicular to the direction of the gas flow unless the chondrule before melting has a peculiar shape as windmill. Fig. 11 shows the deformation of a rotating droplet in gas flow in a three-dimensional view. The rotation axis is set to be perpendicular to the direction of the gas flow. We use $\mu_d =$

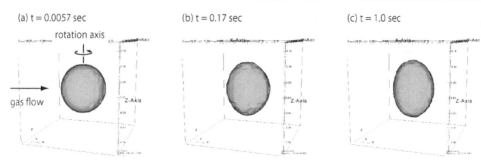

Fig. 11. Three-dimensional view of a rotating molten droplet exposed to a high-velocity gas flow. The object shows the external shape of the droplet (iso-surface of the color function of $\phi = 0.5$). The gas flow comes from the left side (arrow). The rotation axis of the droplet is perpendicular to the direction of the gas flow. After $t = 1.0$ sec, the droplet shape becomes a prolate. We use $\mu_d = 10^3$ poise, $p_{fm} = 10^4$ dyn cm^{-2}, $\omega = 100$ rad s^{-1}, and $r_0 = 1$ mm.

10^3 poise, $p_{fm} = 10^4$ dyn cm^{-2}, $\omega = 100$ rad s^{-1}, and $r_0 = 1$ mm. It is found that the droplet elongates in a direction of the rotation axis as the time goes by. Fig. 12 shows the time variation of the axial ratios b/a (solid) and c/b (dashed). The major axis a corresponds to the droplet radius in a direction of the gas flow, so the decrease of b/a means the droplet elongation. The axial ratio b/a reaches a steady value of 0.76 after 1 sec. The axial ratio c/b is kept at a constant value of ≈ 0.95 during the calculation, which means that two droplet radius perpendicular to the rotation axis is almost uniform. The droplet shape at the steady state is prolate, in other words, a rugby-ball-like shape.

4.4 Origin of prolate chondrule

Why did the droplet shape become prolate? The reason, of course, is due to the droplet rotation. If there is no rotation on the droplet, its shape is only affected by the gas which comes from the fixed direction (see Fig. 13a). In this case, the droplet shape becomes disk-like (oblate) shape because only one axis, which corresponds to the direction of the gas flow, becomes shorter than the other two axes (Sekiya et al., 2003). In contrast, let us consider the case that the droplet is rotating. If the rotation period is much shorter than the viscous deformation timescale, the gas flow averaged during one rotation period can be considered to be axis-symmetrical about the rotation axis (see Fig. 13b). Therefore, the droplet shrinks due to the axis-symmetrical gas flow along directions perpendicular to the rotation axis and becomes prolate if the averaged gas ram pressure is strong enough to overcome the centrifugal force.

Doi (Doi, 2011) derived the analytic solution of deformation of a rotating droplet in gas flow in a case that the gas flow can be approximated as axis-symmetrical around the rotation axis as shown in Fig. 13(b). He considered that the droplet radius is given by $r(\theta) = r_0 + r_1(\theta)$, where r_0 is the unperturbed droplet radius and r_1 is the deviation from a perfect sphere. θ is the angle between the position (the origin is the center of the droplet) and the rotation axis. According to his solution, the droplet deformation is given by

$$\frac{r_1(\theta)}{r_0} = \frac{W_e}{12}\left(\frac{19}{20} - R\right) P_2(\cos\theta), \qquad (35)$$

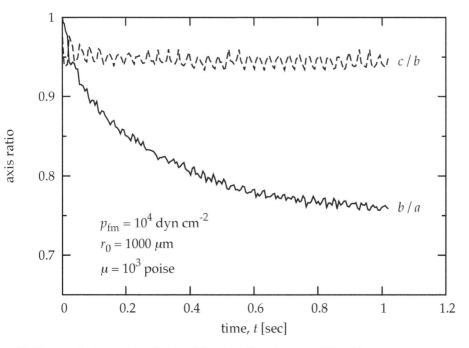

Fig. 12. Time evolutions of axial ratios b/a and c/b in the case of Fig. 11.

(a) no rotation   (b) with rotation of viscous droplet

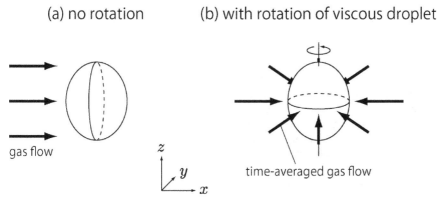

Fig. 13. The reason why the rotating droplet exposed to the gas flow is deformed to a prolate shape is illustrated. (a) If the droplet does not rotate, it is deformed only by the effect of the gas ram pressure. (b) If the droplet rotates much faster than the deformation due to the gas flow, the time-averaged gas flow can be approximated as axis-symmetrical around the rotation axis.

where W_e (Weber number) is the ratio of the ram pressure of the gas flow to the surface tension of the droplet defined as

$$W_e = \frac{p_{fm} r_0}{\gamma_s}, \tag{36}$$

R is the ratio of the centrifugal force to the ram pressure defined as

$$R = \frac{\rho_d r_0^2 \omega^2}{p_{fm}}, \tag{37}$$

ω is the angular velocity of the rotation, and $P_l(\cos\theta)$ is Legendre polynomials. This solution is applicable under the assumption of $r_1 \ll r_0$. Eq. (35) shows that the particle radius becomes the maximum at $\theta = 0$, and minimum at $\theta = \pi/2$. $R = 19/20$ is a critical value for the droplet shape to be prolate ($R < 19/20$) or oblate ($R > 19/20$). The droplet shape is sphere when $R = 19/20$ because the ram pressure balances with the centrifugal force.

Fig. 14 shows the droplet shape as functions of the Weber number W_e and the normalized centrifugal force R using Eq. (35). $R = 19/20$ (vertical dashed line) is a critical value for the droplet shape to be prolate ($R < 19/20$) or oblate ($R > 19/20$). In the prolate region, the axial ratio b/a is less than unity for $W_e > 0$ as shown by contours, but $c/b = 1$. On the other hand, in the oblate region, the axial ratio c/b is less than unity for $W_e > 0$, but $b/a = 1$. As W_e increases, the degree of deformation increases as shown in decrease of axial ratio b/a or c/b. The blue and red regions show ranges of axial ratios of group-A spherical chondrules and group-B prolate chondrules, respectively. We carried out the hydrodynamics simulations for a wide range of parameters and displayed on this diagram by symbols. It is found that the hydrodynamics simulation results show a good agreement with the analytic solution for a wide range of W_e and R.

Let us consider the shape of chondrule expected from the shock-wave heating model. Adopting ram pressure of the gas flow of $p_{fm} = 10^4$ dyn cm^{-2} and the radius of chondrule of $r_0 = 1$ mm, we obtain $W_e = 2.5$ for $\gamma_s = 400$ erg cm^{-2}. According to Eq. (34), we evaluate $R = 0.06$ for $f = 0.01$. The evaluated value of R is smaller than the critical value of $19/20$, so the expected droplet shape is prolate. In addition, the axial ratio b/a comes into a range of group-B prolate chondrules (see Fig. 14). This suggests that the origin of group-B prolate chondrules can be explained by the shock-wave heating model. Of course, it should be noted that the shock-wave heating model does not reproduce the group-B prolate chondrules for arbitrary conditions because W_e and R depend on many factors, e.g., p_{fm}, r_0, and f. Namely, it is possible that different shock conditions produce different chondrule shapes, even out of the range of group-A or -B. This fact, on the contrary, indicates that the chondrule shapes constrain shock conditions suitable for formation of these chondrules. The data of three-dimensional chondrule shapes measured by Tsuchiyama et al. (Tsuchiyama et al., 2003) is definitely valuable, however, the number of samples is twenty at most. We need more data to constrain the chondrule formation mechanism from their three-dimensional shapes.

5. Fragmentation

5.1 Direct fragmentation

When the droplet size is too large for the surface tension to keep the droplet shape against the gas ram pressure, the fragmentation will occur. Fig. 15 shows the three-dimensional views of

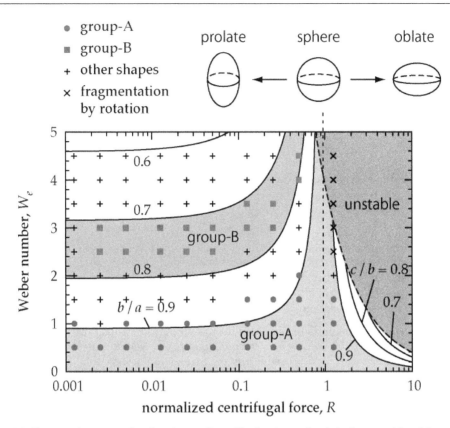

Fig. 14. Shapes of rotating droplets in gas flow. The horizontal axis is the centrifugal force normalized by ram pressure of the gas flow R. The vertical axis is the Weber number W_e. $R = 19/20$ (vertical dashed line) is a critical value for the droplet shape to be prolate ($R < 19/20$) or oblate ($R > 19/20$). Solid lines are contours of axial ratios of b/a ($R < 19/20$) or c/b ($R > 19/20$). A ranges of axial ratios of chondrules are shown by colored regions for group-A spherical chondrules (blue) and for group-B prolate chondrules (red), respectively. Symbols are results of hydrodynamics simulations (see legends in figure). Grayed region shows a condition in which the droplet will be fragmented by rapid rotation.

the break-up droplet. The droplet radius is $r_0 = 2$ cm, which corresponds to $W_e = 20$. The gas flow comes from the left side of the view along the x-axis. It is found that the droplet shape is deformed as the time goes by (panels (a) and (b)), and leads to fragmentation (panel (c)). The parent droplet breaks up to many smaller pieces.

Susa & Nakamoto (Susa & Nakamoto, 2002) suggested that the fragmentation of the droplets in high-velocity gas flow limits the sizes of chondrules (upper limit). They considered the balance between the surface tension and the inhomogeneity of the ram pressure acting on the droplet surface, and derived the maximum size of molten silicate dust particles above which the droplet would be destroyed by the ram pressure of the gas flow using an order of magnitude estimation. In their estimation, they adopted the experimental data in which

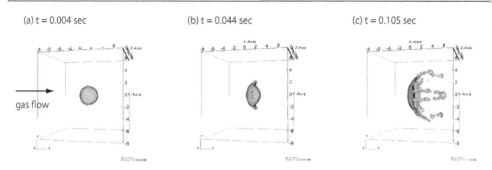

Fig. 15. Three-dimensional view of the fragmentation of molten droplet. We use $\mu_d = 1$ poise, $p_{fm} = 4000$ dyn cm^{-2}, and $r_0 = 2$ cm. The calculation was performed on a $100 \times 100 \times 100$ grid.

the droplets suddenly exposed to the gas flow fragment for $W_e > \sim 6$ (Bronshten, 1983, p.96). This results into the fragmentation of droplet for $r_0 > \sim 6$ mm if we adopt our calculation conditions: $p_{fm} = 4000$ dyn cm^{-2} and $\gamma_s = 400$ dyn cm^{-1}. Our hydrodynamics simulations agree with the criterion for fragmentation.

5.2 Fragmentation via cavitation

Fig. 16 shows the internal pressure inside the droplet for various droplet sizes: $r_0 = 3, 4$, and 5 mm from panels (a) to (c). We use $\mu_d = 1.3$ poise and $p_{fm} = 4000$ dyn cm^{-2}. These droplets reach steady states, so their hydrodynamics do not change significantly after these panels. We found a high pressure region at the front of the droplet, and low pressure regions at centers of eddies in all cases. The high pressure is due to the ram pressure of the gas flow. The low pressure in eddy is clearly due to the non-linear effect caused by the advection term in Eq. (2). Surprisingly, the pressure in eddy decreases to almost zero in panels (b) and (c). In the "zero"-pressure region, the vaporization (or boiling) of the liquid would take place because the vapor pressure of the liquid exceeds the internal pressure. This phenomenon is well known as cavitation. We did not take into account the cavitation in our simulations, so no vaporization occurred in the calculation. If the cavitation was taken into consideration, the eddies are no longer maintained because of the cavitation, which would cause the fragmentation of the droplet.

Miura & Nakamoto (Miura & Nakamoto, 2007) proposed the condition for the "zero"-pressure region to appear by considering the balance between the centrifugal force and the pressure gradient force around eddies as $\rho_d v_{circ}^2 / r_{eddy} \approx p / r_{eddy}$, where v_{circ} is the fluid velocity around the eddy, r_{eddy} is the radius of the eddy, and p is the pressure inside the droplet. Substituting $p = 2\gamma_s / r_0$ from the Young-Laplace equation and $v_{circ} \approx v_{max} = 0.112 p_{fm} r_0 / \mu_d$ (Sekiya et al., 2003), we obtain

$$r_{0,cav} \approx \left(\frac{2\gamma_s \mu_d^2}{0.112^2 \rho_d p_{fm}^2} \right)^{1/3}. \tag{38}$$

This equation gives the critical radius of the droplet above which the cavitation takes place in the center of the eddy. We obtain $r_{0,cav} = 1.3$ mm for the calculation condition. In our hydrodynamic simulations, we observed the "zero"-pressure region for $r_0 = 4$ mm or larger.

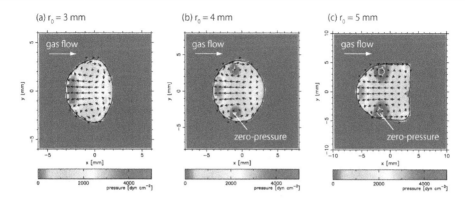

Fig. 16. Internal pressure inside droplet for different droplet radius r_0: (a) 3 mm, (b) 4 mm, and (c) 5 mm. The pressure at a region surrounded by a white line decreases to almost zero by the eddy. We use $\mu_d = 1.3$ poise and $p_{fm} = 4000$ dyn cm^{-2}.

The inconsistency of cavitation criterion between hydrodynamics simulation and Eq. (38) might come from the fact that we substitute the linear solution into v_{circ}. The Sekiya's solution did not take into account the non-linear term in the Navier-Stokes equation. On the other hand, the cavitation would be caused by the non-linear effect. The substitution of the linear solution into the non-linear phenomenon might be a reason for the inconsistency. However, Eq. (38) provide us an insight of the cavitation criterion qualitatively.

5.3 Comparison with chondrule properties

It was found from the chondrule size distribution (see Fig. 3) that chondrules larger than a few mm in radius are very rare. The origin of the chondrule size distribution has been considered as some size-sorting process prior to chondrule formation in the early solar gas disk (Teitler et al., 2010, and references therein). On the other hand, in the framework of the shock-wave heating model, the upper limit of chondrule sizes can be explained by the fragmentation of a molten chondrule in high-velocity gas flow. The criterion of fragmentation is given by $W_e = p_{fm} r_0 / \gamma_s \approx 6$. Since the ram pressure of the gas flow is typically $p_{fm} \approx 10^3 - 10^5$ dyncm^{-2}, we obtain the upper limit of chondrule sizes as $r_{max} \approx 0.2 - 20$ mm. This is consistent with the fact that chondrules larger than a few mm in radius are very rare.

In addition, our hydrodynamics simulations show a new pathway to the fragmentation by cavitation. The cavitation takes place for $W_e < 6$ if viscosity of the molten chondrule is small. The viscosity decreases rapidly as temperature of the droplet increases. This suggests the following tendency: chondrules that experienced higher maximum temperature during melting have smaller sizes that that experienced lower maximum temperature. On the other hand, the data obtained by Nelson & Rubin (Nelson & Rubin, 2002) showed the tendency opposite from our prediction. They considered the reason of the difference in mean sizes among chondrule textural types being due mainly to parent-body chondrule-fragmentation events and not to chondrule-formation processes in the solar nebula. Therefore, to date, there is no evidence regarding the dependence of chondrule sizes on the maximum temperature. The relation between the chondrule sizes and the maximum temperature should be investigated in the future.

How about the distribution of sizes smaller than the maximum one? Kadono and his colleagues carried out aerodynamic liquid dispersion experiments using shock tube (Kadono & Arakawa, 2005; Kadono et al., 2008). They showed that the size distributions of dispersed droplets are represented by an exponential form and similar form to that of chondrules. In their experimental setup, the gas pressure is too high to approximate the gas flow around the droplet as free molecular flow. We carried out the hydrodynamics simulations of droplet dispersion and showed that the size distribution of dispersed droplets is similar to the Kadono's experiments (Yasuda et al., 2009). These results suggest that the shock-wave heating model accounts for not only the maximum size of chondrules but also their size distribution below the maximum size.

In addition, we recognized a new interesting phenomenon relating to the chondrule formation: the droplets dispersed from the parent droplet collide each other. A set of droplets after collision will fuse together into one droplet if the viscosities are low. In contrary, if the set of droplets solidifies before complete fusion, it will have a strange morphology that is composed of two or more chondrules adhered together. This is known as compound chondrules and has been observed in chondritic meteorites in actuality. The abundance of compound chondrules relative to single chondrules is about a few percents at most (Akaki & Nakamura, 2005; Gooding & Keil, 1981; Wasson et al., 1995). The abundance sounds rare, however, this is much higher comparing with the collision probability of chondrules in the early solar gas disk, where number density of chondrules is quite low (Gooding & Keil, 1981; Sekiya & Nakamura, 1996). In the case of collisions among dispersed droplets, a high collision probability is expected because the local number density is high enough behind the parent droplet (Miura, Yasuda & Nakamoto, 2008; Yasuda et al., 2009). The fragmentation of a droplet in the shock-wave heating model might account for the origin of compound chondrules.

6. Conclusion

To conclude, hydrodynamics behaviors of a droplet in space environment are key processes to understand the formation of primitive materials in meteorites. We modeled its three-dimensional hydrodynamics in a hypervelocity gas flow. Our numerical code based on the CIP method properly simulated the deformation, internal flow, and fragmentation of the droplet. We found that these hydrodynamics results accounted for many physical properties of chondrules.

7. References

Akaki, T. & Nakamura, T. (2005). Formation processes of compound chondrules in cv3 carbonaceous chondrites: Constraints from oxygen isotope ratios and major element concentrations., *Geochim. Cosmochim. Acta* 69: 2907–2929.

Amelin, T., Krot, A. N., Hutcheon, I. D. & Ulyanov, A. A. (2002). Lead isotopic ages of chondrules and calcium-aluminum-rich inclusions, *Science* 297: 1678–1683.

Amelin, Y. & Krot, A. (2007). Pb isotopic age of the allende chondrules, *Meteorit. Planet. Sci.* 42: 1321–1335.

Blander, M., Planner, H., Keil, K., Nelson, L. & Richardson, N. (1976). The origin of chondrules: experimental investigation of metastable liquids in the system mg2sio4-sio2, *Geochimica et Cosmochimica Acta* 40(8): 889 – 892, IN1–IN2, 893–896.

URL: *http://www.sciencedirect.com/science/article/B6V66-48C8H7W-BK/2/bdd79a0a3820 afc4d06ac02bdc7cfaa7*

rackbill, J. U., Kothe, D. B. & Zemach, C. (1992). A continuum method for modeling surface tension, *Journal of Computational Physics* 100(2): 335 – 354.
URL: *http://www.sciencedirect.com/science/article/pii/002199919290240Y*

ronshten, V. A. (1983). *Physics of Meteoric Phenomena*, Dordrecht: Reidel.

handrasekhar, S. (1965). The stability of a rotating liquid drop, *Proceedings of the Royal Society of London. Ser. A, Mathematical and Physical Sciences* 286: 1–26.

iesla, F. J. & Hood, L. L. (2002). The nebula shock wave model for chondrule formation: Shock processing in a particle-gas suspension, *Icarus* 158: 281–293.

iesla, F. J., Hood, L. L. & Weidenschilling, S. J. (2004). Evaluating planetesimal bow shocks as sites for chondrule formation, *Meteorit. Planet. Sci.* 39: 1809–1821.

esch, S. J. & Jr., H. C. C. (2002). A model of the thermal processing of particles in solar nebula shocks: Application to the cooling rates of chondrules, *Meteorit. Planet. Sci.* 37: 183–207.

oi, M. (2011). *Formation of cosmic spherule: chemical analysis and theory for shapes, compositions, and textures*, PhD thesis, Tokyo Institute of Technology.

redriksson, K. & Ringwood, A. (1963). Origin of meteoritic chondrules, *Geochimica et Cosmochimica Acta* 27(6): 639 – 641.
URL: *http://www.sciencedirect.com/science/article/B6V66-48C8FST-5/2/baf33fee9f7f0a8d0 a3ef0a92ac38cf7*

ooding, J. L. & Keil, K. (1981). Relative abundances of chondrule primary textural types in ordinary chondrites and their bearing on conditions of chondrule formation, *Meteoritics* 16: 17–43.

larold C. Connolly, J. & Hewins, R. H. (1995). Chondrules as products of dust collisions with totally molten droplets within a dust-rich nebular environment: An experimental investigation, *Geochimica et Cosmochimica Acta* 59: 3231–3246.

layashi, C. K., Nakazawa, K. & Nakagawa, Y. (1985). *Formation of the solar system*, Protostars and Planets II, Univ. of Arizona Press, Tucson, pp. 1100–1153.

lewins, R. H. & Radomsky, P. M. (1990). Temperature conditions for chondrule formation, *Meteoritics* 25: 309–318.

lollenbach, D. & McKee, C. F. (1979). Molecular formation and infrared emission in fast interstellar shocks. i. physical proceses, *Astrophys. J.* 41: 555–592.

lood, L. L. (1998). Thermal processing of chondrule precursors in planetesimal bow shocks, *Meteorit. Planet. Sci.* 33: 97–108.

lood, L. L. & Horanyi, M. (1991). Gas dynamic heating of chondrule precursor grains in the solar nebula, *Icarus* 93: 259–269.

lood, L. L. & Horanyi, M. (1993). The nebular shock wave model for chondrule formation - one-dimensional calculations, *Icarus* 106: 179–189.

lughes, D. W. (1978). A disaggregation and thin section analysis of the size and mass distribution of the chondrules in the bjurbi̊s̃e and chainpur meteorites, *Earth and Planetary Science Letters* 38(2): 391 – 400.
URL: *http://www.sciencedirect.com/science/article/pii/0012821X78901139*

ida, A., Nakamoto, T., Susa, H. & Nakagawa, Y. (2001). A shock heating model for chondrule formation in a protoplanetary disk, *Icarus* 153: 430–450.

Jones, R. H., Lee, T., Jr., H. C. C., Love, S. G. & Shang, H. (2000). *Formation of chondrules and CAIs: Theory vs. observation*, Protostars and Planets IV, Univ. of Arizona Press, Tucson. pp. 927–962.

Jones, R. H. & Lofgren, G. E. (1993). A comparison of feo-rich, porphyritic olivine chondrules in unequilibrated chondrites and experimental analogues, *Meteoritics* 28: 213–221.

Kadono, T. & Arakawa, M. (2005). Breakup of liquids by high velocity flow and size distribution of chondrules, *Icarus* 173: 295–299.

Kadono, T., Arakawa, M. & Kouchi, A. (2008). Size distributions of chondrules and dispersed droplets caused by liquid breakup: An application to shock wave conditions in the solar nebula, *Icarus* 197: 621–626.

Kato, T., Nakamoto, T. & Miura, H. (2006). Maximal size of chondrules in shock wave heating model: Stripping of liquid surface in a hypersonic rarefied gas flow, *Meteorit. Planet. Sci.* 41: 49–65.

Landau, L. D. & Lifshitz, E. M. (1987). *Fluid Mechanics, Course of Theoretical Physics*, Vol. 6, 2nd edn, Elsevier/Butterworth/Heinemann, Oxford, UK.

Lofgren, G. & Russell, W. J. (1986). Dynamic crystallization of chondrule melts of porphyritic and radial pyroxene composition, *Geochim. Cosmochim. Acta* 50: 1715–1726.

Miura, H. & Nakamoto, T. (2006). Shock-wave heating model for chondrule formation: Prevention of isotopic fractionation, *Astrophys. J.* 651: 1272–1295.

Miura, H. & Nakamoto, T. (2007). Shock-wave heating model for chondrule formation: Hydrodynamic simulation of molten droplets exposed to gas flows, *Icarus* 188: 246–265.

Miura, H., Nakamoto, T. & Doi, M. (2008). Origin of three-dimensional shapes of chondrules. i: Hydrodynamics simulations of rotating droplet exposed to high-velocity rarefied gas flow, *Icarus* 197: 269–281.

Miura, H., Nakamoto, T. & Susa, H. (2002). A shock-wave heating model for chondrule formation: Effects of evaporation and gas flows on silicate particles, *Icarus* 160: 258–270.

Miura, H., Yasuda, S. & Nakamoto, T. (2008). Fragment-collision model for compound chondrule formation: Estimation of collision probability, *Icarus* 194: 811–821.

Morris, M. A. & Desch, S. J. (2010). Thermal histories of chondrules in solar nebula shocks, *Astrophys. J.* 722: 1474–1494.

Morris, M. A., Desch, S. J. & Ciesla, F. J. (2009). Cooling of dense gas by H_2o line emission and an assessment of its effects in chondrule-forming shocks, *Astrophys. J.* 691: 320–331.

Nagashima, K., Tsukamoto, K., Satoh, H., Kobatake, H. & Dold, P. (2006). Reproduction of chondrules from levitated, hypercooled melts, *J. Crys. Growth* 293: 193–197.

Nakagawa, Y., Sekiya, M. & Hayashi, C. (1986). Settling and growth of dust particles in a laminar phase of a low-mass solar nebula, *Icarus* 67: 375–390.

Nakamura, T., Tanaka, R., Yabe, T. & Takizawa, K. (2001). Exactly conservative semi-lagrangian scheme for multi-dimensional hyperbolic equations with directional splitting technique, *Journal of Computational Physics* 174(1): 171 – 207. URL: *http://www.sciencedirect.com/science/article/pii/S0021999101968883*

Nelson, L. S., Blander, M., Skaggs, S. R. & Keil, K. (1972). Use of a co_2 laser to prepare chondrule-like spherules from supercooled molten oxide and silicate droplets, *Earth Planet. Sci. Lett.* 14: 338–344.

Nelson, V. E. & Rubin, A. E. (2002). Size-frequency distributions of chondrules and chondrule fragments in ll3 chondrites: Implications for parent-body fragmentation of chondrules, *Meteorit. Planet. Sci.* 37: 1361–1376.

Radomsky, P. M. & Hewins, R. H. (1990). Formation conditions of pyroxene-olivine and magnesian olivine chondrules, *Geochim. Cosmochim. Acta* 54: 3475–3490.

Rubin, A. E. (1989). Size-frequency distributions of chondrules in co3 chondrites, *Meteoritics* 24: 179–189.

Rubin, A. E. & Grossman, J. N. (1987). Size-frequency-distributions of eh3 chondrules, *Meteoritics* 22: 237–251.

Rubin, A. E. & Keil, K. (1984). Size-distributions of chondrule types in the inman and allan hills a77011 l3 chondrites, *Meteoritics* 19: 135–143.

Ruzmaikina, T. V. & Ip, W. H. (1994). Chondrule formation in radiative shock, *Icarus* 112: 430–447.

Sekiya, M. & Nakamura, T. (1996). Condition for the formation of the compound chondrules in the solar nebula, *Proc. NIPR Symp. Antarct. Meteorites* 9: 208–217.

Sekiya, M., Uesugi, M. & Nakamoto, T. (2003). Flow in a liquid sphere moving with a hypersonic velocity in a rarefied gas—an analytic solution of linearized equations, *Prog. Theor. Phys.* 109: 717–728.

Srivastava, A., Inatomi, Y., Tsukamoto, K., Maki, T. & Miura, H. (2010). In situ visualization of crystallization inside high temperature silicate melts, *J. Appl. Phys.* 107: 114907.

Susa, H. & Nakamoto, T. (2002). On the maximal size of chondrules in shock wave heating model, *Astrophys. J.* 564: L57–L60.

Teitler, S. A., Paque, J. M., Cuzzi, J. N. & Hogan, R. C. (2010). Statistical tests of chondrule sorting, *Meteorit. Planet. Sci.* .

Tsuchiyama, A. & Nagahara, H. (1981-12). Effects of precooling thermal history and cooling rate on the texture of chondrules: A preliminary report, *Memoirs of National Institute of Polar Research. Special issue* 20: 175–192.
    URL: *http://ci.nii.ac.jp/naid/110000009441/*

Tsuchiyama, A., Nagahara, H. & Kushiro, I. (1980). Experimental reproduction of textures of chondrules, *Earth Planet. Sci. Lett.* 48: 155–165.

Tsuchiyama, A., Osada, Y., Nakano, T. & Uesugi, K. (2004). Experimental reproduction of classic barred olivine chondrules: Open-system behavior of chondrule formation, *Geochim. Cosmochim. Acta* 68: 653–672.

Tsuchiyama, A., Shigeyoshi, R., Kawabata, T., Nakano, T., Uesugi, K. & Shirono, S. (2003). Three-dimensional structures of chondrules and their high-speed rotation, *Lunar Planet. Sci.* 34: 1271–1272.

Tsukamoto, K., Satoh, H., Takamura, Y. & Kuribayashi, K. (1999). A new approach for the formation of olivine-chondrules by aero-acoustic levitation, *Antarct. Meteorites* 24: 179–181.

Uesugi, M., Akaki, T., Sekiya, M. & Nakamura, T. (2005). Motion of iron sulfide inclusions inside a shock-melted chondrule, *Meteorit. Planet. Sci.* 40: 1103–1114.

Uesugi, M., Sekiya, M. & Nakamoto, T. (2003). Deformation and internal flow of a chondrule-precursor molten sphere in a shocked nebular gas, *Earth Planets Space* 55: 493–507.

Wasson, J. T., Alexander, N. K., Lee, M. S. & Rubin, A. E. (1995). Compound chondrules, *Geochim. Cosmochim. Acta* 59: 1847–1869.

Wood, J. A. (1984). On the formation of meteoritic chondrules by aerodynamic drag heating in the solar nebula, *Earth Planet. Sci. Lett.* 70: 11–26.

Xiao, F., Yabe, T. & Ito, T. (1996). Constructing oscillation preventing scheme for advection eqution by rational function, *Comp. Phys. Comm.* 93: 1–12.

Yabe, T. & Aoki, T. (1991). A universal solver for hyperbolic equations by cubic-polynomial interpolation i. one-dimensional solver, *Comp. Phys. Comm.* 66: 219–232.

Yabe, T. & Wang, P. Y. (1991). Unified numerical procedure for compressible and incompressible fluid, *J. Phys. Soc. Jpn.* 60: 2105–2108.

Yabe, T., Xiao, F. & Utsumi, T. (2001). The constrained interpolation profile method for multiphase analysis., *J. Comp. Phys.* 169: 556–593.

Yasuda, S., Miura, H. & Nakamoto, T. (2009). Compound chondrule formation in the shock-wave heating model: Three-dimensional hydrodynamics simulation of the disruption of a partially-molten dust particle, *Icarus* 204: 303–315.

7

Rotational Dynamics of Nonpolar and Dipolar Molecules in Polar and Binary Solvent Mixtures

Sanjeev R. Inamdar
Laser Spectroscopy Programme,
Department of Physics,
Karnatak University, Dharwad
India

. Introduction

'he absorption of photons by a molecule leads to its excitation. An electronically excited molecule can lose its energy by emission of ultraviolet, visible, infrared radiation or by collision with the surrounding matter. Luminescence is thus the emission of photons from xcited electronic energy levels of molecules. The energy difference between the initial and he final electronic states is emitted as fluorescence or phosphorescence (Lakowicz, 2006). 'luorescence is a spin-allowed radiative transition between two states of the same nultiplicity (e.g., $S_1 \rightarrow S_0$) whereas; phosphorescence is a spin-forbidden radiative transition etween two states of different multiplicity (e.g., $T_1 \rightarrow S_0$). 'he mechanisms by which electronically excited molecules relax to ground state are given y the Jablonski diagram as shown in Fig. 1. The absorption of a photon takes a molecule rom ground state (singlet state, S_0) to either first excited state (singlet state, S_1) or second

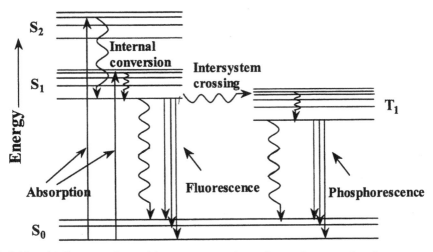

'ig. 1. Jablonski diagram of transitions among various electronic energy levels excited state (S_2). The excited molecule then relaxes to the lowest vibronic level of the first excited state through internal conversion (IC), which generally occurs within 10^{-12} s or less. Since fluorescence lifetimes are typically near 10^{-8} s, IC is generally complete prior to emission. Now it can relax from the singlet excited state to the ground state via three mechanisms. First by emitting a photon (radiative process), second without emitting photon (nonradiative mechanism) and third it goes to a triplet state (T_1) by intersystem crossing (ISC) which also is a nonradiative process. The transition from triplet (T_1) to ground singlet state is forbidden and hence is a very slow process relative to fluorescence. Emission from T_1 is called phosphorescence and generally is shifted to longer wavelength relative to the fluorescence.

In fluorescence spectroscopy the observed spectral intensity is a function of two variables the excitation wavelength (λ_{ex}) and the emission wavelength (λ_{em}). The fluorescence property of a compound is conventionally studied by examining both the excitation spectrum and the emission spectrum. The intensity vs. wavelength plot of the fluorescence spectrum obtained is characteristic of a fluorophore and sensitive to its local surrounding environment. It is consequentially used to probe structure of the local environment. Generally, the wavelength of maximum fluorescence intensity is shifted to longer wavelength relative to the wavelength of its absorption maximum. The difference between these two wavelengths, known as Stokes' shift, arises because of the relaxation from the initially excited state to the 'ground' vibronic level of S_1 which involves a loss of energy. Further loss of energy is due to the transitions from S_1 to higher vibrational levels of the ground state S_0. The Stokes' shift further increases because of general solvent effects. The energy difference between the absorption maximum (v_a) and the emission maximum (v_f) is given by Lippert equation (Birks, 1970) in which the energy difference (v_a-v_f) of a fluorophore as a function of the refractive index (n) and dielectric constant (ε) of the solvent is related as

$$v_a - v_f \approx \frac{2}{hc}\left[\frac{\varepsilon-1}{2\varepsilon+1} - \frac{n^2-1}{2n^2+1}\right]\frac{(\mu^*-\mu)^2}{a^3} + const \tag{1}$$

where h is the Planck's constant, c the velocity of light and a is the radius of the cavity in which the fluorophore resides. Also, μ and μ^* are the ground and excited state dipole moments, respectively.

Fluorescence emission is generally independent of excitation wavelength. This is because of the rapid relaxation to the lowest vibrational level of S_1 prior to emission, irrespective of excitation to any higher electronic and vibrational levels. Excitation on the extreme red edge of the absorption spectrum frequently results in a red-shifted emission. The red-shift occurs because red-edge excitation selects those fluorophores which are more strongly interacting with the solvent (solvation dynamics) (Demchenko, 2002). The red-edge effect can also be thought as ground state heterogeneity, which is common in most complex systems like a probe distribution in microheterogeneous media. In the case of ground state heterogeneity or the presence of multiple species in the ground state, the fluorescence emission spectrum is dependent on the excitation wavelength and the fluorescence excitation spectrum is dependent on the emission wavelength. Also fluorescence excitation spectrum observed for a given emission wavelength differs from that of the absorption spectrum for heterogeneous system. The large spectral width of the emission spectrum compared to absorption spectral width is also due to the presence of multiple species in the excited state. Fluorescence emission spectrum is generally a mirror image of the absorption spectrum (S_0 to S_1 transition).

1.1 Steady-state and time resolved fluorescence

Fluorescence measurements can be broadly classified into two types of measurements: steady-state and time-resolved. Steady-state measurements, the most common type, are those performed with constant illumination and observation. The sample is illuminated with a continuous beam of light, and the intensity or emission spectrum is recorded as function of wavelength. When the sample is first exposed to light steady state is reached almost immediately. Because of the ns timescale of fluorescence, most measurements employ steady-state method. The second type of measurement is time-resolved method which is used for measuring intensity decays or anisotropy decays. For these measurements the sample is exposed to a pulse of light, where the pulse width is typically shorter than the decay time of the sample. The intensity decay is recorded with a high-speed detection system that permits the intensity or anisotropy to be measured on the ns timescale.

1.2 Fluorescence anisotropy

The photoselection of fluorescent probe by polarized light offers the opportunity to study some relevant processes occurring at molecular level in heterogeneous systems. The fluorescence, emitted from the samples excited with polarized light, is also polarized. This polarization is due to the photoselection of the fluorophores according to their orientation relative to the direction of the polarized excitation. This photoselection is proportional to the square of the cosine of the angle between the absorption dipole of the fluorophore and the axis of polarization of the excitation light. The orientational anisotropic distribution of the excited fluorophore population relaxes by rotational diffusion of the fluorophores and excitation energy transfer to the surrounding acceptor molecule. The polarized fluorescence emission becomes depolarized by such processes. The fluorescence anisotropy measurements reveal the average angular displacement of the fluorophore, which occurs between absorption and subsequent emission of a photon. The degree of polarization, P, and steady state fluorescence anisotropy r, are thus respectively given by equations (Lakowicz, 2006)

$$P = \frac{I_{||} - I_{\perp}}{I_{||} + I_{\perp}} \tag{2}$$

$$r = \frac{I_{||} - I_{\perp}}{I_{||} + 2I_{\perp}} \tag{3}$$

where $I_{||}$ and $I_{\perp}$ represent the fluorescence intensities when the orientation of the emission polarizer is parallel and perpendicular to the orientation of the excitation polarizer, respectively. The fluorescence anisotropy (r) is a measure of the average depolarization during the lifetime of the excited fluorophore under steady-state conditions. A steady-state observation is simply an average of the time-resolved phenomena over the intensity decay of the sample. But the time resolved measurements of fluorescence anisotropy using ultrafast polarized excitation source (laser) give an insight into the time dependent depolarization. The time dependent fluorescence anisotropy decay, $r(t)$, is defined as

$$r(t) = \frac{I_{||}(t) - I_{\perp}(t)}{I_{||}(t) + 2I_{\perp}(t)} \tag{4}$$

where $I_{||}(t)$ and $I_{\perp}(t)$ are the fluorescence intensity decays collected with the polarization of the emission polarizer maintained parallel and perpendicular to the polarization of the excitation source, respectively. For a fluorophore in a sample solvent, the fluorescence depolarization is simply due to rotational motion of the excited fluorophore and the decay parameters depend on the size and shape of the fluorophore. For spherical fluorophores, the anisotropy decay is a single exponential with a single rotational correlation time and is given by (Lakowicz, 2006)

$$r(t) = r_0 \exp(-t / \tau_r) \tag{5}$$

where r_0 is the initial anisotropy (anisotropy at time t=0 or anisotropy observed in the absence of any depolarizing processes) and τ_r is the rotational correlation time. The initial anisotropy r_0 is related to the angle (θ) between the absorption and emission dipoles of the fluorophore under study as

$$r_0 = \frac{2}{5}\left(\frac{3\cos^2(\theta) - 1}{2} \right) \tag{6}$$

where the value r_0 can vary between 0.4 and –0.2 as the angle (θ) varies between 0^0 and 90^0 respectively. The rotational correlation times τ_r of the fluorophore is governed by the viscosity (η), temperature (T) of the solution and the molecular volume (V) of the fluorophore. This is given by Stokes-Einstein relation (Fleming, 1986) as shown below:

$$\tau_r = \frac{\eta V}{kT} \tag{7}$$

where k is the Boltzmann constant.

The relation between the steady-state anisotropy (r), initial anisotropy (r_0), rotational correlation time (τ_r) and fluorescence lifetime (τ_f) is given by Perrin equation as follows (Lackowicz, 1983)

$$\frac{r_0}{r} = 1 + \frac{\tau_f}{\tau_r} \tag{8}$$

The Perrin equation is very useful in obtaining the correlation time without the measurement of polarization dependent fluorescence decays [$I_{||}(t)$ and $I_{\perp}(t)$]. The theory developed for more complicated shapes of the fluorophore show that a maximum of five exponentials are enough to explain the fluorescence anisotropy decay (Steiner, 1991).

2. Introduction to rotational dynamics

Understanding solute-solvent interaction has been of great relevance in physico-chemical processes due to the importance of these interactions in determining properties such as chemical reaction yield and kinetics or the ability to isolate one compound from another. Interactions between the solutes and their surrounding solvent molecules are difficult to resolve because, unlike in solids, the spatial relationship between the molecules are not fixed on time scales that can be accessed using structural measurements such as X-ray diffraction or multidimensional NMR spectrometry. Intermolecular interactions in the liquid phase are more complex than those in gas phase because of their characteristic strength, the property that gives rise to the liquid phase and at the same time prevents a simple statistical description of collisional interactions from providing adequate insight (Fleming, 1986).

Regardless of almost three and a half decades of continuous investigation, the details of solute-solvent interactions, particularly in polar solvent systems, remain to be understood in detail. Most investigations of intermolecular interactions in solution have used a "probe" molecule present at low concentration in neat or binary solvent systems. Typically, a short pulse of light is shone to establish some non-equilibrium condition in the ensemble of probe molecules, with the object of the experiment being to monitor the return to equilibrium. These studies have included fluorescence lifetime, molecular reorientation (Eisenthal, 1975; Shank and Ippen, 1975; von Jena and Lessing, 1979a; Sanders and Wirth, 1983; Templeton et al., 1985; Blanchard and Wirth, 1986; Templeton and Kenney-Wallace, 1986; Blanchard, 1987, 1988, 1989; Blanchard and Cihal, 1988; Hartman et al., 1991; Srivastava and Doraiswamy, 1995; Imeshev and Khundkar, 1995; Dutt, et al., 1995; Chandrashekhar et al., 1995; Levitus et al., 1995; Backer et al., 1996; Biasutti et al., 1996; Horng et al., 1997; Hartman et al., 1997; Laitinen et al., 1997; Singh, 2000; Dutt and Raman, 2001; Gustavsson et al., 2003; Dutt and Ghanty, 2004; Kubinyi et al., 2006), vibrational relaxation (Heilweil et al., 1986, 1987, 1989; Lingle Jr. et al., 1990; Anfinrud et al., 1990; Elsaesser and Kaiser, 1991; Hambir et al., 1993; Jiang and Blanchard, 1994a & b, 1995; McCarthy and Blanchard, 1995, 1996) and time-delayed fluorescence Stokes shift (Shapiro and Winn, 1980; Maroncelli and Fleming, 1987; Huppert et al. 1989, 1990; Chapman et al., 1990; Wagener and Richert, 1991; Fee et al., 1991; Jarzeba et al., 1991; Yip et al., 1993; Fee and Maroncelli, 1994; Inamdar et al., 1995) measurements. Of these, molecular reorientation of molecules in solution has been an important experimental and theoretical concept for probing the nature of liquids and the interactions of solvents with molecules. This has proven to be among the most useful because of the combined generality of the effect and the well-developed theoretical framework for the interpretation of the experimental data (Debye, 1929; Perrin, 1936; Chuang and Eisethal, 1972; Hu and Zwanzig, 1974; Youngren and Acrivos, 1975; Zwanzig and Harrison, 1985). Though, the effect of solute-solvent interactions on the rotational motion of a probe molecule in solution has been extensively studied, these interactions are generally described as friction to probe rotational motion and can be classified into three types. The first category includes short-range repulsive forces, which dominate intermolecular dynamics during molecular collisions. These interactions are present in all liquids and lead to viscous dissipation, which is well described by hydrodynamic theories (Fleming, 1986). The second category includes long-range electrostatic interactions between a charged or dipolar probe and polar solvent molecules. As the solute turns, the induced solvent polarization can lag behind rotation of the probe, creating a torque, which systematically reduces the rate of rotational diffusion. This effect, termed dielectric friction, arises from the same type of correlated motions of solvent molecules, which is responsible for the time dependent Stokes' shift (TDSS) dynamics of fluorescent probes (van der Zwan and Hynes, 1985; Barbara and Jarzeba, 1990; Maroncelli, 1993). The third category includes specific solute-solvent interactions. Hydrogen bonding is probably the most frequently encountered example of this kind. Strong hydrogen bonds will lead to the formation of solute-solvent complexes of well-defined stoichiometry. These new, larger species can persist in solution for fairly long times and will rotate more slowly than the bare solute. Formation and breakage of weak H binds occurring on time scales faster than probe rotation will provide a channel for rotational energy dissipation giving rise to additional friction.

The theoretical interest in the study of rotational reorientation kinetics of molecules in liquids arises from the fact that it provides information about the intermolecular interaction in the condensed phase. However, the theoretical modeling of molecular reorientation in liquids and its correlation with experimental data is still far from satisfactory. Thus far, two kinds of approaches have been employed in understanding the rotational dynamics. In the first approach, binary collision approximation has been used to explain the rotational dynamics. With this approach, kinetic theory model for rotational relaxation has been employed for rough sphere fluids (Widom, 1960; Rider and Fixman, 1972; Chandler, 1974) and for smooth convex bodies (Evans et al., 1982; Evans and Evans, 1984; Evans, 1988). Evans model along with Enskog equation for viscosity has been employed to express rotational reorientation time (τ_r) as a function of the solvent viscosity. However, explaining rotational dynamics from such a molecular point of view is severely constrained on account of multibody interaction in a fluid. For real systems the quantitative predictions can be made about the variation of τ_r with solvent viscosity. The second approach is the macroscopic approach of understanding the rotational dynamics, where the solvent is assumed to be a structureless continuum and the rotational motion of solutes is considered Markovian or diffusional. A considerable degree of success on the rotational dynamics arises from the Stokes-Einstein-Debye (SED) hydrodynamic theory, which forms the basis of understanding molecular rotations of medium sized molecules (few hundred Å^3 volumes) in liquids (Einstein, 1906; Debye, 1929; Stokes, 1956), according to which the rotational reorientation time (τ_r) of a solute molecule is proportional to its volume (V), bulk viscosity (η) of the solvent and inversely related to its temperature (T).

Rotational dynamics of number of nonpolar and polar solutes have been carried out in homologous series of polar and nonpolar solvents. In general, the experimentally measured reorientation times of most of the nonpolar probes could be described by the SED theory with slip boundary condition. In some cases the reorientation times were found to be faster than predicted by the slip boundary condition, a situation termed as subslip behavior. However, for a given probe in a homologous series of solvents (alkanes or alcohols) the normalized reorientation times (i.e., reorientation times at unit viscosity) decrease as the size of the solvent increased. In other words, the reorientation times did not scale linearly with solvent viscosity. This behavior, known as the size effects, could not be explained with SED theory. Another observation, which the SED theory failed to explain, is that the experimentally measured reorientation times of nonpolar probes are faster in alcohols than in alkanes of similar viscosity. To explain the observed size effects two quasihydrodynamics theories have been used. The first one is a relatively old theory proposed by Geirer and Wirtz (GW) (1953), which takes into account both the size of the solute as well as that of the solvent while calculating the boundary condition. This theory visualizes the solvent to be made up of concentric shells of spherical particles surrounding the spherical probe molecule at the center. Each shell moves at a constant angular velocity and the velocity of successive shells decreases with the distance from the surface of the probe molecule, as though the flow between the shells is laminar. As the shell number increases, i.e., at large distances, the angular velocity vanishes. Although, the GW theory is successful in predicting the observed ize effects in a qualitative way, it could not explain the faster rotation of nonpolar probes in lcohols compared to alkanes. The second relatively new quasihydrodynamic theory was roposed by Dote, Kivelson and Schwartz (DKS) (1981). The DKS theory not only takes into onsideration the relative sizes of the solvent and the probe but also the cavities or free paces created by the solvent around the probe molecule. If the size of the solute is omparable to the free volumes of the solvent, the coupling between the solute and the olvent will become weak which results in reduced friction experienced by the probe molecule.

)n the other hand, rotational dynamics of small and medium sized polar solutes dissolved 1 polar solvents experiences more friction than predicted by the hydrodynamic theories. his 'additional friction' is attributed to the solute-solvent hydrogen bonding. The first and he oldest concept of dielectric friction invoked by chemists is the 'solvent-berg' model, in vhich it is assumed that there is a solute-solvent interaction causing increase in the volume f the solute. Such an enhancement of the volume automatically causes the molecule to otate slower. However, reservations against such an explanation have also been expressed Chuang and Eisenthal, 1972; Horng et al., 1997). Objections to this kind of interpretations rise from the fact that in bulk solution, the solvent molecules are expected to exchange solute-solvent hydrogen bonding dynamics) on a much faster time scale compared to the otational dynamics. Later, the slower reorientation times of polar molecules in polar olvents have been interpreted using dielectric friction theories (Phillips et al., 1985; Dutt et l., 1990; Alavi et al., 1991b,c; Dutt and Raman, 2001; Gustavsson et al., 2003). Dielectric riction on a rotating solute arises because the polar molecule embedded in a dielectric nedium polarizes the surrounding dielectric. As the solute tries to rotate, the polarization of he medium cannot instantaneously keep in phase with the new orientation of the probe molecule and this lag exerts a retarding force on the probe molecule, giving rise to rotational lielectric friction. Although molecular theories of dielectric friction are available, at present hese theories are difficult to apply because they require some knowledge of the ntermolecular potential or some unavailable properties of the solvent. Continuum theories •ffer advantages of simplicity and the calculation of molecular friction in terms of easily ccessible bulk properties of the solvent.

'he SED theory has been found to describe the rotational dynamics of medium sized nolecules fairly well when the coupling between the solute and solvent is purely nechanical or hydrodynamic in nature. It is documented that the SED model correctly •redicts the linear dependence of the rotational reorientation times on the solvent viscosity or polar and cationic dyes dissolved in polar and non polar solvents (Chuang and :isenthal, 1971; Fleming et al., 1976; 1977; Porter et al., 1977; Moog et al., 1982; Spears and :ramer, 1978; Millar et al., 1979; von Jena and Lessing, 1979a, b; 1981; Rice and Kenney-Vallace, 1980; Waldeck and Fleming, 1981; Dutt et al., 1990; Alavi et al., 1991a, b, c; :rishnamurthy et al., 1993; Dutt et al., 1998) that have been interpreted using dielectric iction theories. The dielectric friction can be modeled using continuum theories of Nee-'wanzig (NZ) (Nee and Zwanzig, 1970), which treats the solute as a point dipole rotating in spherical cavity, Alavi-Waldeck (AW) (Alavi and Waldeck, 1991b; 1993) model which is an •xtension of the NZ theory where the solute is treated as a distribution of charges instead of •oint dipole and the semiempirical approach of van der Zwan and Hynes (vdZH) (van der 'wan and Hynes, 1985) in which fluorescence Stokes shift of the solute in a given solvent is elated to dielectric friction. Conversely, the results of neutral and nonpolar solutes deviate significantly from the hydrodynamic predictions at higher viscosities (Waldeck et al., 1982 Canonica et al., 1985; Phillips et al., 1985; Courtney et al., 1986; Ben Amotz and Drake, 1988 Roy and Doraiswamy, 1993; Williams et al., 1994; Jiang and Blanchard, 1994; Anderton and Kauffman, 1994; Brocklehurst and Young, 1995; Benzler and Luther, 1997; Dutt et al., 1999 Ito et al., 2000; Inamdar et al., 2006). These probes rotate much faster than predicted by the SED theory with stick boundary condition and are described by either slip boundary condition or by quasihydrodynamic theories. Slip boundary condition (Hu and Zwanzig, 1974) assumes the solute-solvent coupling parameter to be less than unity, contrary to the stick boundary condition. Quasihydrodynamic theories of Gierer and Wirtz (GW) (Gierer and Wirtz, 1953) and Dote, Kivelson and Schwartz (DKS) (Dote, Kivelson and Schwartz, 1981) attempt to improve upon SED theory by taking into consideration not only the size of the solute but also that of the solvent molecule, thereby modifying the boundary conditions. It has been argued (Ben Amotz and Drake, 1988; Roy and Doraiswamy, 1993) that as the size of the solute molecule becomes much larger than the size of the solvent molecule, the observed reorientation times approach the SED theory with the stick boundary condition. Based on the above description, we have chosen two kinds of solutes categorized as nonpolar and polar to study their rotational reorientation dynamics in nonpolar, polar and binary mixtures of solvents. In the first case, where the nonpolar probes embedded in polar or nonpolar solvents to examine the influence of solute to solvent size ratio and the shape of the solute on the friction experienced by the probe molecule which in turn enables to test the validity of hydrodynamic and quasihydrodynamic theories. The friction experienced by these probes is purely hydrodynamic or mechanical in nature since it is dominated by short-range repulsive forces. Polar probes used in charged polar solvents with an intention of understanding how the long-range electrostatic interactions between the solute and the solvent, which are charge-dipole or dipole-dipole in nature, influence the rotational dynamics of the probe molecules. Dielectric friction on a rotating solute arises because of the polar molecule entrenched in a dielectric medium polarizes the surrounding dielectric. As the solute tries to rotate, the polarization of the medium cannot instantaneously keep in phase with the new orientation of the probe molecule and this lag exerts a retarding force on the probe molecule, giving rise to rotational dielectric friction.

2.1 Theoretical background

Among the many proposed models for the study of rotational motion, the most commonly employed is the rotational diffusion model outlined by Debye (Debye, 1929), in which the reorientation is assumed to occur in small angular steps. On account of high frequency collisions, a molecule can rotate through a very small angle before undergoing another reorienting collision. The rotational diffusion equation solved to obtain the rotational correlation time τ_r of the density function $\rho(\theta, \phi)$ is given by (Lackowicz, 2006)

$$\frac{\partial \rho}{\partial t} = D \frac{1}{\sin\theta} \frac{\partial}{\partial \theta} \left[\frac{1}{\sin\theta} \frac{\partial \rho}{\partial \theta} + \frac{1}{\sin^2\theta} \frac{\partial^2 \rho}{\partial \phi^2} \right] \qquad (9)$$

where D is the rotational diffusion coefficient. For spherical particles ρ satisfies the form $C_1(t) Y_{l,m}(\theta, \phi)$ in isotropic liquids, where $Y_{l,m}(\theta, \phi)$ are the Legendre polynomials and the coefficient $C_1(t)$ is essentially the same as the correlation function. Substitution of $\rho = C_1(t) Y_{l,m}(\theta, \phi)$ gives an ordinary differential equation for C as

$$\frac{dC_1}{dt} = -Dl(l+1)C_1 \tag{10}$$

This implies that the correlation function decays exponentially, $e^{-t/\eta}$ and the correlation time $\tau_l = [l(l+1)D]^{-1}$. In fluorescence depolarization experiments, one measures the anisotropy decay which is $l=2$ correlation and hence $\tau_r = (6D)^{-1}$.

The rotational diffusion co-efficient of a solute is given by the Stokes-Einstein model Lakowicz, 2006) as

$$D = \frac{kT}{\zeta} \tag{11}$$

where ζ is the friction coefficient and kT is the thermal energy. It is this friction, which is of great importance in theoretical as well as experimental studies. A molecule rotating in liquid experiences friction on account of its continuous interaction with its neighbors and the desire to understand has been a motivating force in carrying the experimental measurements of rotational reorientation in liquids.

2.1.1 Hydrodynamic theory

Mechanical friction on a rotating solute in solvent is computed employing hydrodynamic theory by treating the solute as a smooth sphere rotating in a continuum fluid, which is characterized by a shear viscosity. If 'a' is the radius of the molecule and 'η' the viscosity of the liquid, then according to Stokes law (Stokes, 1956)

$$\zeta = 8\pi a^3 \eta \tag{12}$$

Eqn. (11) reduces to

$$D = \frac{kT}{8\pi\eta a^3} \tag{13}$$

The rotational correlation time (τ_r) is given by

$$\tau_r = \frac{1}{6D} = \frac{\zeta}{6kT} \tag{14}$$

substitution of Eqn. (12) in (14) gives

$$\tau_r = \frac{\eta V}{kT} \tag{15}$$

where V is the molecular volume. The most widely used Stokes-Einstein-Debye (SED) hydrodynamic equation for the description of rotational dynamics of spherical molecule is given by

$$\tau_r = \frac{\eta V}{kT} + \tau_0 \tag{16}$$

where τ_0 is the rotational reorientation time at zero viscosity. It is known that spherical approximation embedded in a SED is glossy in error and the shape of the probes is however, more important. In reality, the exact shape of the solute molecule is need not be a spherical and there is a necessary to include a parameter, which should describe the exact shape of nonspherical probes. Hence, the equation for nonspherical molecule proposed by Perrin (Perrin, 1936) is given as follows

$$\tau_r = \frac{\eta V}{kT}(fC)$$
(17)

where f is shape factor and is well specified, C is the boundary condition parameter dependent strongly on solute, solvent and concentration. The shapes of the solute molecules are usually incorporated into the model by treating them as either symmetric or asymmetric ellipsoids. For nonspherical molecules, $f > 1$ and the magnitude of deviation of f from unity describes the degree of the nonspherical nature of the solute molecule. C, signifies the extent of coupling between the solute and the solvent and is known as the boundary condition parameter (Barbara and Jarzeba, 1990). In the two limiting cases of hydrodynamic stick and slip for a nonspherical molecule, the value of C follows the inequality, $0 < C \le 1$ and the exact value of C is determined by the axial ratio of the probe.

It is observed that the experimentally measured rotational reorientation times of number of the nonpolar solutes (Waldeck et al., 1982; Canonica et al., 1985; Phillips et al., 1985; Courtney et al., 1986; Ben Amotz and Drake, 1988; Roy and Doraiswamy, 1993; Williams et al., 1994; Jiang and Blanchard, 1994; Anderton and Kauffman, 1994; Brocklehurst and Young, 1995; Benzler and Luther, 1997; Dutt et al., 1999; Ito et al., 2000; Inamdar et al., 2006) could be described by the SED theory with slip boundary condition (subslip behavior). For a homologous series of solvents such as alcohols or alkanes, the normalized reorientation times decreased as the size of the solvent is increased. In other words, the reorientation times did not scale linearly with solvent viscosity.

2.1.2 Quasihydrodynamic theories

While the SED hydrodynamic theory takes only the size of the solute molecule into account leaving solvent size aside, one needs to consider the size of the solute as well as solvent molecules. Quasihydrodynamic theories consider these and modify the boundary condition accordingly. To explain such observation of size effects, two quasihydrodynamic theories by Gierer and Wirtz (GW) and Dote, Kivelson and Schwartz (DKS) have been used.

i.   Gierer and Wirtz theory (GW)

The first and the relatively old theory proposed by Girer and Wirtz (GW) in 1953, takes into account both the size of the solute as well as that of the solvent while calculating the boundary condition. It visualizes the solvent to be made up of concentric shells of spherical particles surrounding the spherical probe molecule at the center. Each shell moves at a constant angular velocity and the velocity of successive shells decreases with the distance from the surface of the probe molecule, as though the flow between the shells is laminar. As the shell number increase, i.e., at large distances, the angular velocity vanishes. The angular velocity ω_1 of the first solvation shell is related to the angular velocity ω_0 of the probe molecule by means of a sticking factor σ.

$$\omega_1 = \sigma\omega_0$$
(18)

When $\sigma = 1$, it gives the stick boundary condition and σ is related to the ratio of the solute to solvent size, as

$$\sigma = \left[1 + 6\left(\frac{V_s}{V_p} \right)^{\frac{1}{3}} C_0 \right]^{-1} \tag{19}$$

where

$$C_0 = \left\{ \frac{6(V_s / V_p)^{1/3}}{\left[1 + 2\left(V_s / V_p\right)^{1/3} \right]^4} + \frac{1}{\left[1 + 4\left(V_s / V_p\right)^{1/3} \right]^3} \right\}^{-1} \tag{20}$$

V_s and V_p are the volumes of the solvent and probe, respectively. The expression for C_{GW} is given by

$$C_{GW} = \sigma C_0 \tag{21}$$

C in Eqn. (17) should be replaced with C_{GW} obtained from Eqn. (21) for calculating the reorientation times with GW theory. When the ratio V_s/V_p is very small C_{GW} reduces to unity and the SED equation with stick boundary condition is obtained.

ii. The Dote-Kivelson-Schwartz theory (DKS)

Although, the GW theory is successful in predicting the observed size effects in a qualitative way, it could not explain the faster rotation of nonpolar probes in alcohols compared alkanes. Hence, the second relatively new quasihydrodynamic theory, was proposed by Dote, Kivelson and Schwartz (DKS) in 1981. This theory not only takes into consideration the relative sizes of the solvent and the probe but also the cavities or free spaces created by the solvent around the probe molecule. If the size of the solute is comparable to the free volumes of the solvents, the coupling between the solute and the solvent will become weak which results in reduced friction experienced by the probe molecule. According to DKS theory the solute-solvent coupling parameter, C_{DKS} is given by (Dote, Kivelson and Schwartz, 1981)

$$C_{DKS} = \left(1 + \gamma / \phi\right)^{-1} \tag{22}$$

where γ / ϕ is the ratio of the free volume available for the solvent to the effective size of the solute molecule, with

$$\gamma = \frac{\Delta V}{V_p} \left[4\left(\frac{V_p}{V_s} \right)^{2/3} + 1 \right], \tag{23}$$

and ϕ is the ratio of the reorientation time predicted by slip hydrodynamics to the stick prediction for the sphere of same volume. ΔV is the smallest volume of free space per solvent molecule and some discretion must be applied while calculating this term (Dutt et al., 1988; Anderton and Kauffman, 1994; Dutt and Rama Krishna, 2000). ΔV is empirically related to the solvent viscosity, the Hilderbrand-Batchinsky parameter B and the isothermal compressibility k_T of the liquid by

$$\Delta V = Bk_T \eta kT \tag{24}$$

Since the Frenkel hole theory and the Hilderbrand treatment of solvent viscosity were developed for regular solutions (Anderton and Kauffman, 1994), Equation (24) may not be a valid measure of the free space per solvent molecule for associative solvents like alcohols and polyalcohols. Hence, for alcohols ΔV is calculated using

$$\Delta V = V_m - V_s \tag{25}$$

where V_m is the solvent molar volume divided by the Avogadro number.

2.1.3 Dielectric friction theories

The simple description of hydrodynamic friction arising out of viscosity of the solvent becomes inadequate when the motion concerning rotations of polar and charged solutes desired to be explained. A polar molecule rotating in a polar solvent experiences hindrance due to dielectric friction (ζ_{DF}), in addition to, the mechanical (ζ_M) or hydrodynamic friction. In general, the dielectric and mechanical contributions to the friction are not separable as they are linked due to electrohydrodynamic coupling (Hubbard and Onsager, 1977; Hubbard, 1978; Dote et al., 1981; Felderhof, 1983; Alavi et al., 1991c; Kumar and Maroncelli, 2000). Despite this nonseparability, it is common to assume that the total friction experienced by the probe molecule is the sum of mechanical and dielectric friction components, i.e.,

$$\zeta_{Total} = \zeta_M + \zeta_{DF} \tag{26}$$

Mechanical friction can be modeled using both hydrodynamic (Debye, 1929) and quasihydrodynamic (Gierer and Wirtz, 1953; Dote et al., 1981) theories, whereas, dielectric friction is modeled using continuum theories.

The earliest research into dielectric effects on molecular rotation took place in the theoretical arena. Initial investigations were closely intertwined with the theories of dielectric dispersion in pure solvents (Titulaer and Deutch, 1974; Bottcher and Bordewijk, 1978; Cole, 1984). Beginning with the first paper to relate the dielectric friction to rotational motion published by Nee and Zwanzig in 1970, a number of studies have made improvements to the Nee-Zwanzig approach (Tjai et al, 1974; Hubbard and Onsager, 1977; Hubbard and Wolynes, 1978; Bordewijk, 1980; McMahon, 1980; Brito and Bordewijk, 1980; Bossis, 1982; Madden and Kivelson, 1982; Felderhof, 1983; Nowak, 1983; van der Zwan and Hynes, 1985; Alavi et al, 1991a,b,c; Alavi and Waldeck, 1993). These have included the electrohydrodynamic treatment which explicitly considers the coupling between the hydrodynamic (viscous) damping and the dielectric friction components.

i. The Nee-Zwanzig theory

Though not the first, the most influential early treatment of rotational dielectric friction was made by Nee and Zwanzig (NZ) (1970). These authors examined rotational dynamics of the same solute/solvent model in the simple continuum (SC) description i.e., they assumed an Onsager type cavity dipole with dipole moment μ and radius a embedded in a dielectric continuum with dispersion $\varepsilon(\omega)$. Motion was assumed to be in the purely-diffusive (or Smoluchowski) limit. Using a boundary condition value calculation of the average reaction field, Nee and Zwanzig obtained their final result linking the dielectric friction contribution in the spherical cavity as

$$\tau_{DF}^{NZ} = \frac{\mu^2}{9a^3kT} \frac{(\varepsilon_\infty + 2)^2(\varepsilon_0 - \varepsilon_\infty)}{(2\varepsilon_0 + \varepsilon_\infty)^2}\tau_D \tag{27}$$

where ε_0, ε_∞ and τ_D are the zero frequency dielectric constant, high-frequency dielectric constant and Debye relaxation time of the solvent, respectively.
If one assumes that the mechanical and dielectric components of friction are separable, then

$$\tau_r^{obs} = \tau_{SED} + \tau_{DF} \tag{28}$$

Therefore, the observed rotational reorientation time (τ_r^{obs}) is given as the sum of reorientation time calculated using SED hydrodynamic theory and dielectric friction theory.

$$\tau_r^{obs} = \frac{\eta VfC}{kT} + \frac{\mu^2}{9a^3kT}\frac{(\varepsilon + 2)^2(\varepsilon_0 - \varepsilon_\infty)\tau_D}{(2\varepsilon_0 + \varepsilon_\infty)^2} \tag{29}$$

It is clear from the above equation that for a given solute molecule, the dielectric friction contribution would be significant in a solvent of low ε and high τ_D. However, if the solute is large, the contribution due to dielectric friction becomes small and the relative contribution to the overall reorientation time further diminishes due to a step increase in the hydrodynamic contribution. Hence, most pronounced contribution due to dielectric friction would be seen in small molecules with large dipole moments especially in solvents of low ε and large τ_D.

i. The van der Zwan-Hynes theory (vdZH)

A semiempirical method for finding dielectric friction proposed by van der Zwan and Hynes (1985), an improvement over the Nee and Zwanzig model, provides a prescription for determining the dielectric friction from the measurements of response of the solute in the solvent of interest. It relates dielectric friction experienced by a solute in a solvent to solvation time, τ_s, and solute Stokes shift, S. According to this theory the dielectric friction is given by (van der Zwan and Hynes, 1985)

$$\tau_{DF} = \frac{\mu^2}{(\Delta\mu)^2}\frac{S\tau_s}{6kT} \tag{30}$$

where $\Delta\mu$ is the difference in dipole moment of the solute in the ground and excited states and

$$S = h\nu_a - h\nu_f \tag{31}$$

where $h\nu_a$ and $h\nu_f$ are the energies of the 0-0 transition for absorption and fluorescence, respectively. The solvation time is approximately related to the solvent longitudinal relaxation time, $\tau_L = \tau_D(\varepsilon_\infty / \varepsilon_0)$ and is relatively independent of the solute properties. Hence, τ_L can be used in place of τ_s in Eqn. (30).
Assuming the separability of the mechanical and dielectric friction components, the rotational reorientation time can be expressed as

$$\tau_r^{obs} = \frac{\eta VfC}{kT} + \frac{\mu^2}{(\Delta\mu)^2}\frac{hc\,\Delta\nu}{6kT}\tau_s \tag{32}$$

where the first term represents the mechanical contribution and the second the dielectric contribution.

iii. The Alavi and Waldeck theory (AW)

Alavi and Waldeck theory (Alavi and Waldeck, 1991a), proposes that it is rather the charge distribution of the solute than the dipole moment that is used to calculate the friction experienced by the solute molecule. Not only the dipole moment of the solute, but also the higher order moments, contribute significantly to the dielectric friction. In other words, molecules having no net dipole moment can also experience dielectric friction. AW theory has been successful compared to NZ and ZH theories in modeling the friction in nonassociative solvents (Dutt and Ghanty, 2003). The expression for the dielectric friction according to this model is given by (Alavi and Waldek, 1991a)

$$\tau_{DF} = P\frac{(\varepsilon_0 - 1)}{(2\varepsilon_0 + 1)^2}\tau_D \tag{33}$$

where

$$P = \frac{4}{3akT}\sum_{j=1}^{N}\sum_{i=1}^{N}\sum_{L=1}^{L_{max}}\sum_{M=1}^{L}\left(\frac{2L+1}{L+1}\right)\frac{(L-M)!}{(L+M)!}\times$$

$$M^2 q_i q_j \left(\frac{r_i}{a}\right)^L \left(\frac{r_j}{a}\right)^L P_L^M(\cos\theta_i)P_L^M(\cos\theta_j)\cos M\phi_{ji} \tag{34}$$

where $P_L^M(x)$ are the associated Legendre polynomials, a is the cavity radius, N is the number of partial charges, q_i is the partial charge on atom i, whose position is given by (r_i, θ_i, ϕ_i), and $\phi_{ji} = \phi_j - \phi_i$. Although the AW theory too treats solvent as a structureless continuum like the NZ and vdZH theories, it provides a more realistic description of the electronic properties of the solute.

3. Experimental methods

The experimental techniques used for the investigation of rotational reorientation times mainly consist of steady-state fluorescence spectrophotometer and time resolved fluorescence spectrometer employing time correlated single photon counting (TCSPC).

3.1a Steady-state measurements

For vertical excitation, the steady-state fluorescence anisotropy can be expressed as (Dutt et al., 1999; Lakowicz, 1983)

$$<r> = \frac{I_{||} - GI_{\perp}}{I_{||} + 2GI_{\perp}} \tag{35}$$

where $I_{||}$ and $I_{\perp}$ denote the fluorescence intensities parallel and perpendicular polarized components with respect to the polarization of the exciting beam. G (= 1.14) is an instrumental factor that corrects for the polarization bias in the detection system (Inamdar et al., 2006) and is given by

$$G = \frac{I_{HV}}{I_{HH}} \quad (36)$$

where I_{HV} is the fluorescence intensity when the excitation polarizer is kept horizontal and the emission polarizer vertical and I_{HH} is the fluorescence intensity when both the polarizers are kept horizontal.

3.1b Time-resolved fluorescence measurements

The fluorescence lifetimes of all the probes were measured with time correlated single photon counting technique (TCSPC) using equipment described in detail elsewhere (Selvaraju and Ramamurthy, 2004). If the decay of the fluorescence and the decay of the anisotropy are represented by single exponential, then the reorientation time τ_r is given by (Lakowicz, 1983)

$$\tau_r = \frac{\tau_f}{(r_0/<r>-1)} \quad (37)$$

where r_0 is the limiting anisotropy when all the rotational motions are frozen and τ_f is the fluorescence lifetime.

In case of a prolate-ellipsoid model, the parameter f_{stick} is given by (Anderton and Kauffman, 1994)

$$f_{stick} = \frac{2(\rho^2+1)(\rho^2-1)^{3/2}}{3\rho[(2\rho^2-1)\ln\{\rho+(\rho^2-1)^{1/2}\}-\rho(\rho^2-1)^{1/2}]} \quad (38)$$

where ρ is the ratio of major axis (a) to the minor axis (b) of the ellipsoid. This expression is valid for stick boundary condition.

3.2 Fluorescent probes used in the study

Nonpolar probes

A variety of the nonpolar fluorescent probe molecules have been studied extensively in the recent past. Most of the nonpolar probes so far studied have the radii of 2.5 Å to 5.6 Å (Inamdar et al., 2006) and a transition towards stick boundary condition is evident with increase in size of the solute. Most of the medium sized neutral nonpolar molecules rotate faster in alcohols compared to alkanes, which is in contrast to that of smaller neutral solutes. It is also noted that the quasihydrodynamic description is adequate for small solutes of 2-3 Å radius in case of GW theory whereas, the DKS model with experimental value in alcohols fail beyond the solute radius of 4.2 Å. Our earlier work on rotational dynamics of exalite probes E392A ($r = 5.3$ Å), and E398 ($r = 6.0$ Å), yielded striking results (Inamdar et al., 2006), in that, these large probes rotated much faster than slip hydrodynamics and followed subslip trend in alcohols.

The quest to understand the influence of size of solute on rotational dynamics is continued with three nonpolar solutes viz., Exalite 404 (E404), Exalite 417 (E417) and Exalite 428 (E428), which may further fill the gap between the existing data. These probes have an anistropic shape and a dipole emission along their long rod-like backbones. The rod like or cylinder shape is a macromolecular model of great relevance. A number of biopolymers including some polypeptides, proteins, nucleic acids and viruses, under certain conditions exhibit the typical rod-like conformation and their hydrodynamic properties can therefore be analyzed in terms of cylindrical models. Surprisingly, not much is studied about the motion of these highly anisotropic rod-like molecules in liquids, neither experimentally nor by any simulation studies. These exalite dyes have found applications in many areas of research. When pumped by XeCl-excimer laser, Ar+ and Nd:YAG laser, provide tunable lasers in the ultraviolet-blue range (Valenta et al., 1999). E428 has been used to generate circularly polarized light in glassy liquid crystal films (Chen et al., 1999). Exalites are mixed with plastic scintillators (PS) to form new scintillaors, which are for superficial and diagnostic applications (Kirov et al., 1999).

Polar probes

Rotational diffusion of medium-sized molecules provides a useful means to probe solute-solvent interactions and friction. By modeling this friction using various continuum-based theories (NZ, AW and ZH) one can get better insight into the nature of solute-solvent interactions. In order to understand the effect of polar solvents on the reorientational dynamics of the polar solutes, one must unravel the effects of mechanical friction, dielectric friction and specific short-range solute-solvent interactions. To address this issue, rotational dynamics of three polar laser dyes: coumarin 522B (C522B), coumarin 307 (C307) and coumarin 138 (C138) having identical volumes and distinct structures have been carried out in series of alcohols and alkanes. These coumarins are an important class of oxygen heterocycles, which are widespread in plant kingdom and have been extensively used as laser dyes. Their chemical structures can be looked upon as arising out of the fusion of a benzene ring to pyran-2-one, across the 5- and 6-positions in skeleton. In the present coumarins, the two free substituents at 6 and 7 positions, ethylamino and methyl for C307 in comparison with the analogous model substrate C522B wherein, there is no free substituent rather they are joined by ends to obtain piperidino moiety. These two probes are looked upon as polar due to the presence of electron donating amino group and electron withdrawing CF_3 group. In C138, this CF_3 group is replaced by an alkyl group making it less polar compared to C522B and C307.

The rotational diffusion studies of the following two sets of structurally similar molecules dyes: (i) coumarin-440 (C440), coumarin-450 (C450), coumarin 466 (C466) and coumarin-151 (C151) and (ii) fluorescein 27 (F27), fluorescein Na (FNa) and sulforhodamine B (SRB) in binary mixtures of dimethyl sulphoxide + water and propanol + water mixtures, respectively. Among coumarins, C466 possess N-diethyl group at the fourth position whereas, other three dyes possess amino groups at the seventh position in addition to carbonyl group. This structure is expected to influence molecular reorientation due to possible hydrogen bonding with the solvent mixture. The spectroscopic properties of fluorescein dyes are well known with the dyes having applications ranging from dye lasers to tracers in flow visualization and mixing studies. SRB has been used to measure drug-induced cytotoxicity and cell proliferation for large-scale drug-screening applications (Koochesfahani and Dimotakis, 1986; Dahm et al., 1991; Karasso and Mungal, 1997; Voigt, 2005). Both F27 and FNa are neutral polar molecules each containing one $C = O$ group, F-27 has two Cl and FNa has two Na groups. The anionic probe SRB possesses N (C_2H_5), N+ (C_2H_5) groups and sulfonic groups SO_3Na and SO_3^- at positions 3, 6, 4' and 2', respectively.

The laser grade nonpolar probes Exalites (E404, E417 and E428), nonpolar probes (i) coumarin derivatives (C522B, C307 and C138) and (ii) F27, FNa and SRB (all from Exciton Chemical Co., USA) were used as received. For steady-state experiments, all the samples were excited at 375 nm and the emission was monitored from 403-422 nm from alkanes to alcohols for Exalites. All the solvents (Fluka, HPLC grade) were used without further purification. The concentration of all the solutions was kept sufficiently low in order to reduce the effects of self-absorption. All the measurements were performed at 298 K.

3.2.1 Rotational dynamics of non-polar probes

The molecular structures of the non-polar probes exalite 404 (E404), exalite 417 (E417) and exalite 428 (E428) chosen for the study are shown in Fig.2.The absorption and fluorescence spectra of the probes in methanol are shown in Fig.3. These probes are approximated as prolate ellipsoids (Inamdar et al., 2006) with molecular volumes 679, 837 and 1031 $\mathring{A}^3$, respectively, for E404, E417 and E428. The rotational reorientation times (τ_r) calculated using Eqn. (4.43), are tabulated in Table 1 and 2, respectively.

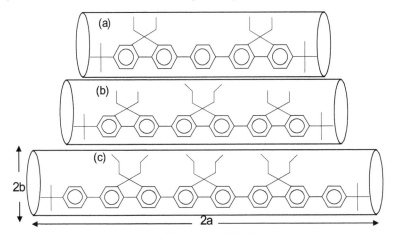

Fig. 2. Molecular structures of (a) E404, (b) E417 and (c) E428

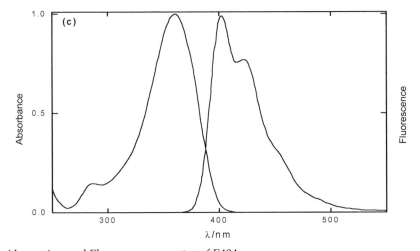

Fig. 3. Absorption and Fluorescence spectra of E404

Solvents	η/ mPa s[a]	E404 τ_r/ ps	E417 τ_r/ ps	E428 τ_r/ ps
Pentane	0.23	114±10	169±14	315±28
Hexane	0.29	153±14	223±18	413±40
Heptane	0.41	199±16	319±22	517±50
Octane	0.52	275±25	401±30	687±65
Nonane	0.66	322±25	523±45	814±78
Decane	0.84	417±35	677±55	1060±101
Dodecane	1.35	634±55	1002±88	1493±141
Tridecane	1.55	718±60	1137±100	1726±155
Pentadecane	2.81	1262±115	1740±120	2261±200
Hexadecane	3.07	1362±113	2080±140	2520±220

[a] Viscosity data is from Inamdar et al., 2006

Table 1. Rotational reorientation times (τ_r) of Exalites in alkanes at 298K

Solvents	η/ mPa s[a]	E404 τ_r/ ps	E417 τ_r/ ps	E428 τ_r/ ps
Methanol	0.55	319±32	675±54	694±63
Ethanol	1.08	494±48	860±70	1244±110
Propanol	1.96	896±73	1153±99	1890±172
Butanol	2.59	1008±89	1710±150	2223±220
Pentanol	3.55	1185±100	1815±155	2750±225
Hexanol	4.59	1514±120	2244±201	3178±300
Heptanol	5.87	2070±182	2363±210	3627±345
Octanol	7.63	2669±255	2859±250	4130±380
Nonanol	9.59	3879±315	3099±292	4541±410
Decanol	11.80	4038±330	3702±340	4800±400

[a] Viscosity data is from Inamdar et al., 2006

Table 2. Rotational reorientation times (τ_r) of Exalites in alcohols at 298K

i. Rotational reorientation times of Exalite 404 (E404)

Fig. 4 gives the plot of τ_r vs η in alkanes and alcohols for E404 shows that τ_r values increase linearly with η both in alkanes and alcohols, following slip hydrodynamic and subslip behavior, respectively. This clearly indicates that the rotational dynamics of E404 follows SED hydrodynamics with slip boundary condition. Further, E404 rotates slower in alkanes compared to alcohols by a factor of 1 to 1.3. It may be recalled that E392A followed SED hydrodynamics near stick limit in alkanes (Inamdar et al., 2006). E404 is larger than E392A by a factor of 1.1, and exhibits an opposite behavior to that of E392A following slip behavior in alkanes. Interestingly, the rotational dynamics of both these probes follow subslip behavior in higher alcohols.

Theoretical justification for this approach is provided by the microfriction theories of Geirer-Wirtz (GW) and Dote-Kivelson-Schwartz (DKS) wherein the solvent size as well as free spaces is taken into account. However, there is a large deviation of experimentally measured reorientation times from those calculated theoretically.

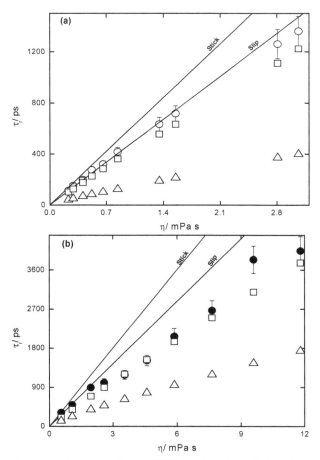

Fig. 4. Plot of rotational reorientation times of E404 as function of viscosity in (a) alkanes and (b) alcohols. The symbols (○,●) represent experimentally measured reorientation times. The stick and slip lines calculated using hydrodynamic theory are represented by solid lines. GW and DKS theories are represented using the symbols △ and   respectively.

i. Rotational reorientation times of Exalite 417 (E417)

The rotational reorientation times of E417 scale linearly with η (Fig. 5) and exhibits subslip behavior in alcohols. A large nonlinearity is observed on increasing solvent viscosity. In alkanes, the rotational reorientation times follow slip hydrodynamic boundary condition, similar to E404. GW theory is unable to explain experimental results while DKS theory is in fairly good agreement with experiment and slip hydrodynamics in case of alkanes.

ii. Rotational reorientation times of Exalite 428 (E428)

E428 is the largest probe studied so far in literature. In alcohols the τ_r values for E428 increase linearly with η from methanol to butanol and follows slip boundary condition, and from pentanol to decanol a large deviation from the linearity is observed resulting in subslip behavior (Fig. 6). However, in alkanes the measured reorientation times, clearly follow slip hydrodynamics up to tridecane, whereas in higher alkanes pentadecane and hexadecane

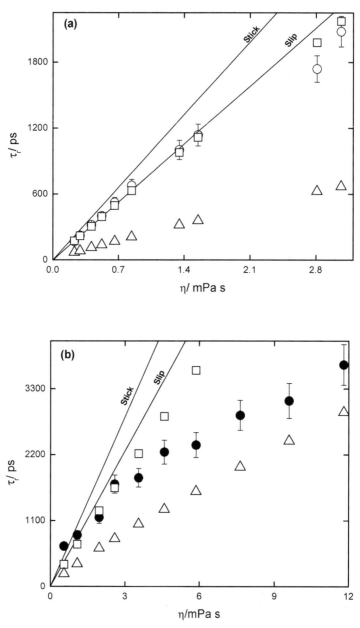

Fig. 5. Plot of rotational reorientation times of E417 as function of viscosity in (a) alkanes and (b) alcohols. The symbols (○,●) represent experimentally measured reorientation times. The stick and slip lines calculated using hydrodynamic theory are represented by solid lines. GW and DKS quasihydrodynamic theories are represented using the symbols △ and   respectively.

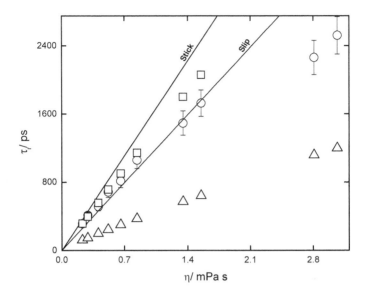

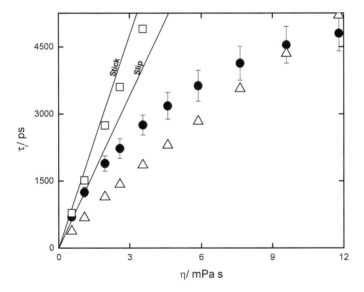

Fig. 6. Plot of rotational reorientation times of E428 as function of viscosity in (a) alkanes and (b) alcohols. The symbols (○,●) represent experimentally measured reorientation times. The stick and slip lines calculated using hydrodynamic theory are represented by solid lines. GW and DKS quasihydrodynamic theories are represented using the symbols Δ and   respectively.

subslip behavior is observed. It is interesting to note that, all the three probes rotate much faster in alcohols compared to alkanes. This can be explained as due to large interstitial gaps that may be formed in the solvent medium and because of the possible elastic nature of the spatial H-bonding network of large alcohol molecules constituting a supramolecular structure. The elasticity of the spatial network is a driving force for solvophobic interaction, which is important for the larger probes. Presumably these exalite molecules will be located mainly in these solvophobic regions. The probe molecules, thus, can rotate more freely in these gaps as they experience reduced friction due to a decreased viscosity at the point of contact. This actual viscosity is highly localized and cannot be measured easily. In such a situation the coupling parameter C can be much smaller than C_{slip} predicted by slip hydrodynamic boundary condition. One of the plausible reasons is also due to the Brownian motion, which results from the fluctuating forces in the liquid, is behind and diffusive process.

Ben-Amotz and Scott (1987) opined that processes, which are slow compared to solvent fluctuations, would see the full spectrum of the fluctuations and thus the shear viscosity of the solvent. For example, the fluctuations in n-alcohols occur roughly on the 100 ps/mPa s time scale – precisely the time scale of the Debye absorption in these solvents. On the other hand, processes, which are extremely fast, do not experience Brownian fluctuating force and are not viscously damped. Thus one expects a reduction in microscopic friction for probe molecules, which diffuse at a rate comparable to or faster than the solvent fluctuations. This is exactly the type of effect, which could explain the faster rotational diffusion of exalites in n-alcohols than in n-alkanes. Further, the subslip behavior observed for these probes in polar solvents indicates the existence the nonhydrodynamic forces and the straightforward relation between the probe size and the nature of their behavior may not be appropriate.

Table 3 and 4 contain selected data for various neutral solute molecules (including exalites), whose rotational times in alkanes and alcohols have been measured experimentally. There are many reports on rotational diffusion of small neutral molecules which follow subslip behavior. Garg and Smyth (1965) have attributed these alcohol molecules to be associated

Solute	Volume (Å^3)	Radius (Å)	Axial ratio	T_r (ps/ cP)	T_{slip} (ps/ cP)	T_{stick} (ps/ cP)	T_r/T_{stick}	References
Biphenyl	152	3.31	2.26	24	20	63	0.38	Bauer et al., 1974
Stilbene	180	3.50	2.82	38	41	94	0.40	Courtney et al., 1986
DPB	208	3.38	3.40	66	71	140	0.47	Waldeck et al., 1982
PTP	224	3.77	3.03	54	59	128	0.42	Philips et al., 1985; Kim et al., 1989
Binapthyl	237	3.84	2.36	51	35	102	0.50	Bowman et al., 1988
PQP	296	4.13	3.77	112	124	226	0.50	Ben Amotz & Scott, 1987
DPA	312	4.15	2.55	78	56	147	0.53	Ben Amotz & Scott, 1987
BMQ	325	4.27	2.65	131	95	241	0.54	Roy and Doraiswamy, 1993
TMQ	359	4.41	2.10	149	71	265	0.56	Roy and Doraiswamy, 1993
DMQ	376	4.48	1.82	194	54	282	0.68	Roy and Doraiswamy, 1993
TMI	429	4.68	2.43	294	145	416	0.71	Roy and Doraiswamy, 1993
E392A	609	5.26	3.59	384	347	435	0.88	Inamdar et al., 2006
QUI	639	5.38	2.09	454	147	554	0.82	Roy and Doraiswamy, 1993
E404	679	5.45	4.21	437	362	601	0.73	Present work
BTBP	733	5.59	3.50	430	269	507	0.85	Ben Amotz & Drake, 1988
E417	837	5.85	5.00	636	623	944	0.67	Present work
E428	1031	6.27	6.18	749	1134	1587	0.47	Present work

Table 3. List of normalized rotational diffusion parameters of neutral nonpolar solutes in alkanes, at 25 ± 5^0 C

Solute	T_r (ps/ cP)	T_r/T_{stick}	References
Stilbene	10	0.10	Courtney et al., 1986
PTP	39	0.30	Ben Amotz & Scott, 1987
Binapthyl	36	0.35	Bowman et al., 1988
DPA	50	0.34	Ben Amotz & Scott, 1987
BMQ	87	0.36	Roy and Doraiswamy, 1993
TMQ	117	0.44	Roy and Doraiswamy, 1993
DMQ	137	0.49	Roy and Doraiswamy, 1993
TMI	194	0.47	Roy and Doraiswamy, 1993
E392A	196	0.45	Inamdar et al., 2006
QUI	436	0.79	Roy and Doraiswamy, 1993
E404	349	0.58	Present work
BTBP	430	0.85	Ben Amotz & Drake, 1988
E417	259	0.27	Present work
E428	357	0.22	Present work

Table 4. List of normalized rotational diffusion parameters of neutral nonpolar solutes in alcohols, at 25 ± 5^0 C

with hydrogen bridges in temporary microcrystalline structures. These structures are in fact not stable, and at a given instant each of these has a finite length. At each instance some hydrogen bonds are ruptured and others are formed.

The first dispersion region is connected with the molecules in these microcrystalline structures. The dielectric relaxation process involves the breaking and reforming of the hydrogen bonds with the orientation of dipole moment, and the rate of breaking off is a determining factor for the relaxation time. In order to check whether there is any dielectric friction on these large nonpolar probes in alcohols, we have also calculated dielectric friction contribution to the rotating probe molecule. The dipole moment values in the excited states were obtained using solvatochormic shift method (Inamdar et al., 2003; Nadaf et al., 2004; Kawski et al., 2005). It is noted that summing up the contribution due to hydrodynamic and dielectric friction will not affect the subslip trend exhibited by the rotational reorientation times. Hence, we attribute this unhindered faster rotation due to strong hydrogen bonding among the solvent molecules leading to supramolecular structures.

There are several reports in literature where the reorientation times of neutral nonpolar solutes have been measured as a function of solute size and the transition from slip to stick hydrodynamics has been observed experimentally. Ben-Amotz and Drake (Ben-Amotz and Drake, 1988) have reported the rotational dynamics of the neutral large sized probe BTBP (V=733 Å^3) in series of alcohols and alkanes, and observed that rotational correlation times followed stick boundary condition. Though, BTBP contain the electronegative groups like -O and –N, which are capable of forming hydrogen bond with any solvent, they attributed, stick condition to its volume which is much larger than that of all the solvent molecules studied. Later, Roy and Doraiswamy (Roy and Doraiswamy, 1993) have studied the rotational dynamics of series of nonpolar solutes, which do not contain any electronegative groups like -O or –N. They observed transition towards the stick boundary condition on increasing the solute size from BMQ (V = 325 Å^3) to QUI (V = 639 Å^3). It is clear from the above two findings that a stick transition arises due to increase in the solute size, when compared to that of the solvent. Thus, one can expect stick or superstick behavior in case of exalites (E404, E417 and E428) as these are larger than QUI by a factor of 1.1, 1.3 and 1.6, respectively. The present situation, where the largest probe E428 follows subslip in alcohols is surprising in the light of above studies. In such a situation the microscopic friction of the solvent molecules reduces well below the macroscopic value, which may result from either dynamic or structural features of the macroscopic solvation environment-giving rise to faster rotation in hydrogen bonding solvents.

On the other hand, rotational reorientation times of these exalite nonpolar probes bequeath interesting results following slip boundary condition in alkanes. It is observed from the Table 5 that there is a difference in slope for the two solvent types. Therefore, it is evident that the rotational reorientation times of these exalites are shorter in alcohols than alkanes of comparable viscosity. This difference is an indication of nonhydrodynamic effects in one or both of the solvents. It is unlikely that nonhydrodynamic behavior resulting from frequency dependence of the solvent friction occurs in alkanes on the 100 ps to 1 ns time scale (Hynes, 1986). These times are much longer than dynamic memory effects in the solvent arising from molecular collisions. These collisional events manifest themselves in the viscoelastic relaxation time, which for an n-alkane is estimated to lie in the subpicosecond to single picosecond time domain (Hynes, 1986).

Solutes	Alcohols	Alkanes
Slope	Slope (ps/ mPa s)	(ps/ mPa s)
E404	349	437
	360*	454*
E417	259	636
	362*	677*
E428	357	749
	510*	901*

* Second entry for solute is a slope of the best fit line made to pass through the origin.

Table 5. Linear regression results of rotational reorientation of exalites in series of alcohols, alkanes and binary mixture

Thus one would expect rotational times to be well described by the SED relation with the appropriate boundary condition and the solute shape factor (Ben Amotz and Scott, 1987) in n-alkanes. The internal mobility also allows the solute molecule to slip better through the surrounding solvent molecules than for a rigid molecular backbone (Alavi et al., 1991b,c). Waldeck et al. (1982) have also argued for the probe DPB, that the slip boundary condition is entirely reasonable for an uncharged nonpolar molecule in nonpolar solvents. E428 is about 5 times larger than DPB and from the Table 3; it is evident that τ_r / τ_{stick} ratio is same for both these probes in alkanes, which suggests the fact that the rotation of these probes can be well explained by slip hydrodynamics. Similarly, the studies of the neutral dye BBOT (Fleming et al., 1977), an approximate prolate top, found that this molecule followed slip boundary condition. It was anticipated that neutrals would not strongly interact with the solvent, and slip boundary condition were thus more appropriate. Others have argued (Porter et al., 1977) that the faster rotation observed for BBOT might also be due to the internal mobility of the dye. This may be one of the possible reasons for the faster rotation observed for the large exalite probes. Both GW and DKS models were tested for a quantitative prediction of τ_r of solutes in alkanes. The GW model predicts very low τ_r values in alkanes as well as in the case of alcohols and fails to satisfactorily explain the observed results. Also, the C values are nearly invariant of the size of the solute. It has een evidenced that the GW theory correctly predicts the observed results for a solute ith ~2.5 Å radius. Therefore, the GW model is adequate for very small solutes that show ubslip behavior, viz., I_2 and NCCCCN (Goulay, 1983). Though, DKS theory is found to be า good agreement with the experimentally observed trend up to decane in case of E404 nd up to nonane for E428, a better agreement is found in alkanes for E417. It has been oted that the rotational reorientation times in alkanes is reproduced quantitatively for olutes with radius up to 4.2 Å only, beyond which the theory tends to show poor greement with experimental values [93]. Our experimental results are indicative of the ıct the DKS theory also holds well even for larger probes up to a radius of 6.3 Å in lkanes and brings out the subtle variations in the observed data.

2.2 Rotational dynamics of polar probes

he rotational dynamics studies using polar solutes in polar solvents have shed lights on oncepts such as dielectric friction and solute-solvent hydrogen bonding. In addition to iscous drag, polar-polar interaction between a polar solute and a polar solvent gives rise to n additional retarding force often termed as dielectric friction. This arises because of the าability of the solvent molecules, encircling the polar solute probe, to rotate synchronously ıith the probe. The result of this effect is the creation of an electric field in the cavity, which xerts a torque opposing the reorientation of the probe molecule. Under such circumstances, าe observed friction, which is proportional to the measured reorientation time, has been xplained as a combination of mechanical and dielectric frictions. However, many xperimental investigations of reorientation dynamics have indicated that there is another ource of drag on a rotating probe molecule due to hydrogen bonding between the solute nd the solvent molecules. A solute molecule can form hydrogen bond with the solvent าolecule depending on the nature of the functional groups on the solute and the solvent ıhich enhances the volume of the probe molecule. This further impedes the rotational าotion and thus the observed reorientation time becomes longer than that observed with าe bare solute molecule.

Aolecular structures of the three coumarin dyes chosen under the category of polar probes re shown in Fig. 7. The reorientation times of C522B, C307 in alcohols and alkanes and :138 in alcohols (Mannekutla et al., 2010) are summarized in Tables 6 and 7. The τ_r values btained in alkanes clearly show that C522B rotates faster compared to C307. In alcohols, it i interesting to note that, the probe C138 rotates faster almost by a factor of 1:2 from ropanol to decanol compared to C522B and C307, respectively. In other words, C138 xperiences a reduced mechanical friction i.e., almost same as C522B and twice as C307 from ropanol to decanol. This is because C307 shows greater interaction owing to its greater olarity.

ig. 7. Molecular structures of (a) C522B, (b) C307 and (c) C138

Solvents	η (mPa s[a])	C522B			C307			C138		
		$<r>$	T_f (ns)	T_r (ps)	$<r>$	T_f (ns)	T_r (ps)	$<r>$	T_f (ns)	T_r (ps)
Methanol	0.55	0.004±0.001	4.87	53±13	0.003±0.001	5.10	42±14	0.003±0.001	3.63	30±10
Ethanol	1.08	0.006±0.001	5.02	83±14	0.006±0.001	5.12	84±14	0.006±0.002	3.63	61±20
Propanol	1.95	0.009±0.001	5.05	126±14	0.011±0.002	5.13	157±29	0.009±0.002	3.64	92±20
Butanol	2.59	0.011±0.002	5.08	156±28	0.016±0.002	5.14	232±29	0.011±0.001	3.64	113±10
Pentanol	3.51	0.013±0.002	5.09	186±29	0.021±0.003	5.14	309±44	0.015±0.003	3.64	156±31
Hexanol	4.57	0.017±0.002	5.09	246±29	0.027±0.002	5.14	405±30	0.020±0.002	3.65	212±21
Heptanol	5.97	0.022±0.003	5.10	323±44	0.034±0.003	5.15	521±46	0.025±0.002	3.65	268±21
Octanol	7.63	0.027±0.002	5.10	403±30	0.041±0.003	5.15	642±47	0.035±0.004	3.68	390±45
Nonanol	9.59	0.033±0.003	5.10	501±46	0.051±0.003	5.18	828±49	0.037±0.002	3.74	422±23
Decanol	11.74	0.042±0.003	5.12	658±47	0.065±0.002	5.18	1104±34	0.047±0.003	3.77	477±35

[a] Viscosity data is from Inamdar et al., 2006

Table 6. Steady-state anisotropy ($<r>$), fluorescence lifetime (T_f) and rotational reorientation time (T_r) of coumarins in alcohols at 298K (the maximum error in the fluorescence life times is less than ±50 ps) (Mannekutla et al., 2010)

Solvents	η/ mPa s[a]	C522B			C307		
		$<r>$	T_f/ ns	T_r/ ps	$<r>$	T_f/ ns	T_r/ ps
Pentane	0.23	-	-	-	0.003±0.001	4.23	35±12
Hexane	0.29	-	-	-	0.003±0.001	4.28	35±12
Heptane	0.41	0.002±0.001	3.98	22±11	0.004±0.001	4.37	48±12
Octane	0.52	0.003±0.001	4.11	34±11	0.004±0.001	4.51	49±12
Nonane	0.66	0.004±0.001	4.17	46±12	0.005±0.001	4.63	63±13
Decane	0.84	0.005±0.001	4.20	58±12	0.006±0.001	4.60	76±13
Dodecane	1.35	0.006±0.001	4.26	70±12	0.007±0.002	4.79	92±26
Tridecane	1.55	0.007±0.001	4.31	83±12	0.008±0.002	4.80	106±27
Pentadecane	2.81	0.009±0.001	4.38	110±12	0.009±0.002	4.80	120±27
Hexadecane	3.07	0.010±0.001	4.52	126±13	0.010±0.002	4.86	135±27

[a] Viscosity data is from Inamdar et al., 2006

Table 7. Steady-state anisotropy ($<r>$), fluorescence lifetime (T_f) and rotational reorientation time (T_r) of coumarins in alkanes at 298K for C522B and C307 (the maximum error in the fluorescence life times is less than ±50 ps) (Mannekutla et al., 2010)

The probes C522B and C138 have shown coincidentally similar interactions. In C138, aminomethyl group being free contributes more to the charge separation through resonance- whereas in C522B this resonance contribution is sluggish, comparatively. However, the presence of -CF$_3$ in C522B increases the charge separation, which leads to better interaction with the hydrogen bonding solvents. Replacement of -CF$_3$ by cyclic alkyl group in C138 would not have any great contribution towards its polarity. Hence, the presence of two different groups with contradicting properties leads to the coincidental similarities in reorientation dynamics of C522B and C138.

he normalized rotational reorientation times (at unit viscosity) are smaller in alkanes ompared to alcohols, which indicates that the probes C522B and C307 rotate faster in lkanes compared to alcohols. The reorientation times of the three probes thus obtained in lcohols follow the trend: $\tau_r^{C307} > \tau_r^{C522B} \geq \tau_r^{C138}$.

ig. 8 gives a typical plot of τ_r vs η for all the three probes in alcohols and in alkanes along ith the stick and slip lines. Note that the experimentally measured reorientation times lie etween slip and stick hydrodynamic in case of alcohols. However, in alkanes we observe, s the size of the solvent molecule becomes equal to and bigger than the size of the solute olecule, the probe molecule experiences a reduced friction. Benzler and Luther (1977) leasured the reorientation time of biphenyl (V=150 Å³) and p-terphenyl (V=221 Å³) in n-lkanes. For biphenyl a nonlinearity was observed in the plot of τ_r vs η from decane and om tetradecane, in case of p-terphenyl. Singh [24] studied reorientation times of the probe eutral red (V=234 Å³) which experienced a reduced friction from tetradecane to exadecane following subslip behavior. C522B (223 Å³) and C307 (217 Å³) have nearly lentical volumes as compared to neutral red and p-terphenyl and thus a similar rotational elaxation in alkanes may be expected.

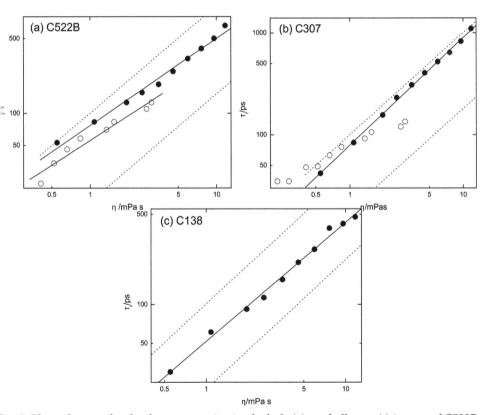

ig. 8. Plots of τ_r vs η for the three coumarins in alcohols ($\circ$), and alkanes ($\bullet$) in case of C522B nd C307

Note that the probes experience reduced friction as the size of the solvent increases. A number of probes have been studied (Phillips et al., 1985; Courtney et al., 1986; Ben Amotz and Drake, 1988; Roy and Doraiswamy, 1993; Williams et al., 1994; Jiang and Blanchard, 1994; Anderton and Kauffman, 1994; Brocklehurst and Young, 1995) in alcohols and alkanes wherein faster rotation of the probe in alcohols is observed compared to alkanes, which has been explained as due to higher free volume in alcohols compared to alkanes with the help of DKS theory. If there were no electrical interaction between the coumarins and alcohols, a faster rotation of the coumarins would have been observed in alcohols compared to alkanes but an opposite trend has been observed that indicates the presence of electrical friction (Dutt and Raman, 2001). Before evaluating the amount of dielectric friction, the contribution due to mechanical friction must be estimated with a reasonable degree of accuracy. SED theory with a slip hydrodynamic boundary condition is often used to calculate the mechanical friction in case of medium-sized solute molecules. However, in the present study the solvent size increases by more than 5 times in alcohols from methanol to decanol. Hence, DKS quasihydrodynamic theory is found to be more appropriate, when size effect is taken into account as compared with GW. Eqn. 25 is used to calculate ΔV in associative solvents like alcohols, because C_{DKS} obtained in this manner gave a better agreement with the experimental results (Hubbard and Onsager, 1977; Anderton and Kauffman, 1996; Dutt et al., 1999; Dutt and Raman, 2001).

In summary, a faster rotation of the probes is observed in case of C522B and C138 in alcohols compared to C307. In spite of the distinct structures, almost similar rotational reorientation times are observed for C522B and C138 in alcohols from propanol to decanol. Further studies of dielectric friction in alcohols, the observed reorientation times of these coumarins could not follow the trend predicted by the theories of Nee-Zwanzig and van der Zwan-Hynes. Dielectric frictions obtained experimentally and theoretically using NZ and ZH theories, do not agree well.

3.2.3 Rotational dynamics of polar probes in binary solvents

Binary mixtures of polar solvents represent an important class of chemical reaction media because their polarity can be controlled through changes in composition. In a binary mixture, altering the composition of one of the ingredients can lead to a change in solubility, polarizability, viscosity and many other static and dynamic properties. Yet, it is often found that the dielectric properties of polar mixtures depart significantly from what one might expect on the basis of ideal mixing. In hydrogen-bonding systems, such as alcohol-water mixtures, intermolecular correlations are strong, and consequently, the dielectric properties of the mixture are usually not simply related to those of the separated components. Recently, the properties of some binary solutions were studied using theoretical calculations and molecular dynamics (MD) simulations (Chandra and Bagchi, 1991; Chandra, 1995; Skaf and Ladanyi, 1996; Day and Patey, 1997; Yoshimori et al., 1998; Laria and Skaf, 1999). The results showed that the dynamical features of binary solutions are very much different from those of neat solutions, and the dynamics can be strongly affected by the properties of the solute probe. The binary mixtures show exotic features which pose interesting challenges to both theoreticians and experimentalists. Amongst them, the extrema observed in the composition dependence of excess viscosity (Qunfang and Yu-Chun, 1999; Pal and Daas, 2000) and the anomalous viscosity dependence of the rotational relaxation time (Beddard et al., 1981) are significant. The anomalous features in the complex systems arise from specific atermolecular interactions due to structural heterogeneities. In DMSO+water mixture, the artial negative charge on the oxygen atom of the dimethyl sulphoxide molecule forms ydrogen bonds with water molecules, giving rise to a non-ideal behavior of the mixture. he non-ideality of mixtures depends on the nature of interaction between the different pecies constituting the mixture. Traube suggested that the anomalous behavior of viscosity 1 binary mixtures arises from the formation of clusters (Traube, 1886). The prominent ydrophilic nature of DMSO renders it capable of forming strong and persistent hydrogen onds with water through its oxygen atom (Safford et al., 1969; Martin and Hanthal, 1975;)e La Torre, 1983; Luzar and Chandler, 1993). This leads to the formation of DMSO-water nolecular aggregates of well-defined geometry which are often responsible for the strong onideal behavior manifested as maxima or minima (Cowie and Toporowski, 1964; Packer nd Tomlinson, 1971; Fox and Whittingham, 1974; Tokuhiro et al., 1974; Gordalla and eidler, 1986; 1991; Kaatze et al., 1989). The largest deviations from the ideal mixing occur round 33% mole of DMSO, thus suggesting the existence of stoichiometrically well defined DMSO:2water complexes. Recently, a number of MD simulations (Vaisman and Berkowitz, 992; Soper and Luzar, 1992; 1996; Luzar and Chandler, 1993; Borin and Skaf, 1998; 1999) nd neutron diffraction experiments have indeed identified the structure of the DMSO:2water complex and linked many of the structural and dynamical features of)MSO water mixtures to the presence of such aggregates. Of late, Borin and Skaf (Borin nd Skaf, 1998; 1999) have found from MD simulations, another distinct type of aggregate onsisting of two DMSO molecules linked by a central water molecule through H-bonding, vhich is expected to be the predominant form of molecular association between DMSO and vater in DMSO-rich mixtures. This H-bonded complex is referred to as 2DMSO:1water ggregate.

'he rotational diffusion studies of the following two sets of structurally similar molecules lyes: coumarin-440 (C440), coumarin-450 (C450), coumarin 466 (C466) and coumarin-151 C151) and fluorescein 27 (F27), fluorescein Na (FNa) and sulforhodamine B (SRB) (Fig. 9) in ·inary mixtures of dimethyl sulphoxide + water and propanol + water mixtures, espectively. Among coumarins, C466 possess N-diethyl group at the fourth position vhereas, other three dyes possess amino groups at the seventh position in addition to arbonyl group. This structure is expected to affect the reorientation times due to the ormation of hydrogen bond with the solvent mixture.

'he photo-physics of fluorescent molecules in solvent mixtures has not been studied as xtensively as those in neat solvents. Thus the structure and structural changes in the olvent environment around the solute in the mixed solvents have not been fully inderstood. It is therefore important to investigate the photophysical characteristics that are nique to the binary solvent mixtures.

)MSO is miscible with water in all proportions and aqueous DMSO solutions are quite nteresting systems, as there exists a nonlinear relationship between the bulk viscosity and he composition of the solvent mixture. In DMSO-water binary mixture, there is a rapid rise n viscosity with a small addition of DMSO to water and viscosity decay profile after the ·ost peak point is gradual. The sharp increase in the viscosity of the binary mixture with ncreasing DMSO concentration may be attributed to significant hydrogen bonding effects ·etween water and DMSO molecules. Beyond around 15% composition of DMSO, there xist two DMSO compositions for which viscosity is same. This dual valuedness should nanifest in reasonable mirror symmetry of the rotational reorientation time (τ_r) about the

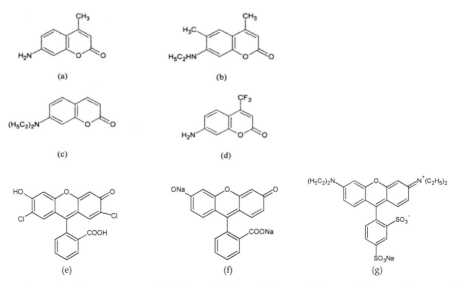

Fig. 9. The molecular structures of (a) C440, (b) C450, (c) C466, (d) C151, (e) F27, (f) FNa and (g) SRB.

viscosity peak point. The viscosity of DMSO is slightly more than twice that of water. At about 40% mole composition of DMSO, the solvent mixture has a maximum value of viscosity of 3.75 m Pas which is 1.87 times that of DMSO and nearly 4 times that of water. From the viscosity profile it may be seen that there are four distinct compositions of DMSO for which the viscosity is nearly the same and as per hydrodynamic theory the friction experienced by a rotating probe molecule is expected to be the same.

Fig. 10 (a and b) represent the variation of τ_r with η along with theoretical profile including the viscous and the dielectric contribution for all the probes, which clearly indicates a non-hydrodynamic behavior. The rotational reorientation time of a solute in a solvent is in a way an index of molecular friction. Experimentally obtained results of all the probes under study show a hairpin profiles bent upwards. The reorientation times gradually increases as a function of viscosity up to the peak viscous value and interestingly these values further increase even after the solvent mixture exhibits reduction in viscosity after the peak value. Thus all the probes exhibit different rotational reorientation values for isoviscous points. Note that, reorientation times are longer in the DMSO region compared to the water rich zone. The studies of the rotational diffusion of the dye molecules in binary solvents showed that the rotational relaxation time does not necessarily scale linearly with viscosity when the solvent composition is changed. These observations have been interpreted as a manifestation of solvent structure on time scales similar to or longer than the time scale of solute rotation or as resulting from a change in the dielectric friction through the solvent mixture. In some cases these observations have been interpreted as a breakdown of the hydrodynamic approximation. The rotational diffusion studies of the dye molecule oxazine 118 in two binary solvent systems as a function of temperature showed a nonlinear dependence of the rotational diffusion on the solvent viscosity when the solvent composition is changed (Williams et al., 1994).

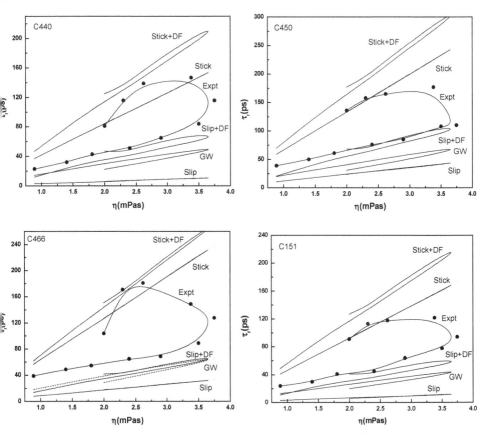

ig. 10. Plot of rotational reorientation time with viscosity along with theoretical profile ncluding the viscous and the dielectric contribution for C440, C450, C466 and C151 probes [namdar et al., 2009)

he linear variation of the τ_r as a function of η from pure water to the composition of the inary mixture when the viscosity reaches its peak is in accordance with the SED theory, hough it does not account for the large curvature in the profile. The theoretical SED stick ne shows a sharp hairpin profile. Incorporation of the dielectric friction contribution ualitatively mimics the observed profile, with the τ_r being slightly larger in the post peak iscosity DMSO rich zone. The fact, that a continuum theory without the consideration of ny molecular features could reproduce the gross features of the observed profile of τ_r vs. η ; noteworthy. The experimentally observed profile bent upwards yields considerably igher τ_r in the DMSO rich zone than the corresponding isoviscous point in the water rich one. This is also reproduced by the theoretical models qualitatively. The pronounced lifference in the rotational reorientation times at the isoviscous points can be explained only n the basis of solvation. It is possible that at the isoviscous points the microstructural eatures in the binary mixture could be different. The dual values of τ_r at isoviscous points n the DMSO rich zone are also due to the contributions of dielectric friction at these two oints being different.

Beddard et al. (1981) reported different rotational relaxation times of the dye cresyl violet in ethanol water mixture by varying the ethanol water composition i.e., at the same viscosity but at different compositions. The observed re-entrance type behavior of the orientational relaxation time when plotted against viscosity could not be explained only in terms of non ideality in viscosity exhibited in a binary mixture. Beddard et al. also reported that the re entrance behavior is strongly dependent on the specific interaction of the solute with the solvents. This is because in a system where solute interacts with few different species in a binary mixture in a different manner, its rotational relaxation will depend more on the composition than on the viscosity of the binary mixture. The role of specific interaction on the orientational dynamics has often been discussed in relation to changing boundary conditions (Fleming, 1986). We find that the orientational relaxation time of the probe molecules when plotted against the solvent viscosity does indeed show re-entrance. Our study here re-affirms that for a solute dissolved in a binary mixture, its rotational relaxation will depend more on the composition than on the viscosity of the binary mixture and thus the re-entrant type behaviour is strongly dependent on the interactions of the solute with the two different species in the solvent.

The rotational dynamics of two kinds of medium sized three dyes-Fluorescein 27(F27) and Fluorescein Na(FNa) (both neutral but polar), and Sulforhodamine B(SRB) (anion) has been studied in binary mixtures comprising of 1-Propanol and water at room temperature using both steady-state and time resolved fluorescence depolarization techniques. Alcohols have both a hydrogen-bonding -OH group and a hydrophobic alkyl group. The latter affects the water structure. The objective in studying two neutral and an anion dyes is to compare and contrast the rotational dynamics as a function of charge. A nonlinear hook-type profile of rotational reorientation times of the probe (τ_r) as a function of viscosity (η) is observed for all three dyes in this binary system, with the rotational reorientation times being longer in organic solvent rich zone, compared to the corresponding isoviscous point in water rich zone. This is attributed to strong hydrogen bonding between the solute and propanol molecules.

The increase in viscosity as 1-propanol is added to water is sharp with the peak value of 2.70 mPa s being reached at about 30% mole composition of 1-propanol. The viscosity of 1-propanol is 1.96 mPa s, the decrease after the post peak point is linear but gradual. The dielectric friction contribution in water, amides, and dipolar aprotics is minimal while it goes on increasing in alcohols (Krishnamurthy et al, 1993).

At isoviscous points there are two different τ_r values and this duality results from different values of dielectric frictions at the isoviscous points (Fig. 11). It is seen that both the neutral dyes F27 and FNa clearly produce the hook-type profile bent upwards and qualitatively mimic the nonhydrodynamic behavior. The reorientation times gradually increase as a function of viscosity up to the peak viscous value. τ_r values decrease after the solvent mixture exhibits a reduction in viscosity after the peak value. Note that the reorientation times are longer in propanol rich region compared to the water rich zone. In case of SRB though it exhibits hook type profile, surprisingly τ_r values longer in water rich zone in the beginning and later probe rotates faster in the intermediate viscous region. In propanol rich zone SRB shows similar τ_r values as those of water rich zone. This may be due to both amino groups of SRB are ethylated and the rotational diffusion of this dye was slightly more rapid than predicted. Theoretical models mimic this trend qualitatively, though GW & DKS models invariably predict a reduced friction and illustrate a hairpin - bending downwards. Thus, these models underestimate the friction experienced by the probe. The dual

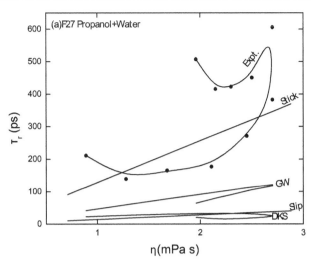

Fig. 11. Plot of rotational reorientation time with viscosity along with theoretical profile including viscous contribution for F27

valuedness of τ_r at isoviscous points near the organic solvent rich zone were attributed to different contributions of dielectric friction at these compositions and to strong hydrogen bonding.

General conclusion and summary

In this article, an attempt has been made to understand solute-solvent interactions in various situations using the powerful fluorescence spectroscopic techniques. The interesting observation of faster rotation of nonpolar probes in alcohols compared to alkanes can be attributed to large interstitial gaps that may be formed in the solvent medium and because of the possible elastic nature of the spatial H-bonding network of large alcohol molecules constituting a supramolecular structure. Presumably the exalite molecules will be located mainly in these solvophobic regions and thus, can rotate more freely in these gaps and experience reduced friction due to decreased viscosity at the point of contact. This actual viscosity is highly localized and cannot be measured easily. In such a situation the coupling parameter C can be much smaller than C_{slip} predicted by slip hydrodynamic boundary condition. Also, the largest probe E428 following subslip trend in alcohols is surprising. In such a situation the microscopic friction of the solvent molecules reduces well below the macroscopic value, which may result from either dynamic or structural features of the macroscopic solvation environment-giving rise to faster rotation in hydrogen bonding solvents. The experimental results indicate that DKS theory also holds well even for larger probes up to a radius of 6.3 Å in alkanes.

In case of polar probes, a faster rotation of the probes is observed for C522B and C138 in alcohols compared to C307. In spite of the distinction in structure a coincidental similar rotational reorientation times is observed in case of C522B and C138 in alcohols from propanol to decanol. Further studies of dielectric friction in alcohols, the observed reorientation times of these coumarins could not follow the trend predicted by the theories of Nee-Zwanzig and van der Zwan-Hynes. Experimentally and theoretically obtained dielectric frictions using NZ and ZH theories, do not agree well.

A nonlinear hook-type profile of rotational reorientation times of the probe as a function of viscosity is observed for all the dyes in binary mixtures, with the rotational reorientation times being longer in organic solvent rich zone, compared to the corresponding isoviscous point in water rich zone. This is attributed to strong hydrogen bonding between the solutes and DMSO or propanol molecules. Theoretical models mimic this trend qualitatively, though GW & DKS models invariably predict a reduced friction and illustrate a hairpin profile bending downwards. Thus they underestimate the friction experienced by the probe. The dual valuedness of τ_r at isoviscous points near the organic solvent rich zone were attributed to different contributions of dielectric friction at these compositions and to strong hydrogen bonding.

In general, the theoretical models: hydrodynamic as well as those based on dielectric friction do not adequately and precisely describe the experimental observations. The theoretical description of solute-solvent interaction to explain the experimental observations is yet to evolve. The failure of the theoretical models, to explain the experimental results quantitatively in specific cases, calls for the formulation of molecular based theories.

4. Acknowledgment

The author acknowledges the encouragement and support of Profs. M.I. Savadatti and B.G. Mulimani. Thanks are also due to Dr. James R.M., K.H. Nagachandra and M.A. Shivkumar for their timely help and financial support from Council of Scientific & Industrial Research and University Grants Commission, New Delhi.

5. References

Alavi, D.S., Hartman, R.S. & Waldeck, D.H., 1991a, A test of continuum models for dielectric friction. Rotational diffusion of phenoxazine dyes in dimethylsulfoxide *J. Chem. Phys.* 94, 4509-20

Alavi, D.S. & Waldeck, D.H., 1991b, Rotational dielectric friction on a generalized charge distribution *J. Chem. Phys.* 94, 6196-6202

Alavi, D.S., Hartman, R.S. & Waldeck, D.H., 1991c, The influence of wave vector dependent dielectric properties on rotational friction. Rotational diffusion of phenoxazine dyes *J. Chem. Phys.* 95, 6770-83

Alavi, D.S. & Waldeck, D.H., 1993, Erratum: Rotational dielectric friction on a generalized charge distribution *J. Chem. Phys.* 98, 3580-82

Anderton, R.M. & Kauffman, J.F., 1994, Temperature-Dependent Rotational Relaxation of Diphenylbutadiene in n-Alcohols: A Test of the Quasihydrodynamic Free Space Model *J. Phys. Chem.* 98, 12117-12124

Anfinrud, P.A., C. Han, T. Lian and R. M. Hochstrasser, 1990, Evolution of the transient vibrational spectrum following short-pulse excitation *J. Phys. Chem.* 94, 1180-84

Backer, S.D., Dutt, G.B., Ameloot, M., Schryver, F.C.D., Müllen, K. & Holtrup, F., 1996, Fluorescence Anisotropy of 2,5,8,11-Tetra-tert-butylperylene and 2,5,10,13-Tetra-tert-butylterrylene in Alkanes and Alcohols *J. Phys. Chem.* 100, 512-518

Barbara, P.F. & Jarzeba, W., 1990, Ultrafast Photochemical Intramolecular Charge and Excited State Solvation in *Adv. in Photochem.* Vol.15, pp. 1-68, Eds: D.H. Volman, G.S. Hammond, K. Gollnick, John Wiley & Sons, Inc.,, USA

Bauer, D.R., I.I. Brauman & Pecora, R., 1974, Molecular reorientation in liquids. Experimental test of hydrodynamic models J.Am.Chem.Soc. *96, 6840-43*

Beddard, G.S., Doust, T. & Hudales, J. 1981, Structural features in ethanol-water mixtures revealed by picoseconds fluorescence anisotropy *Nature* 294,145-46

Ben Amotz, D. & Scott, T.W., 1987, Microscopic frictional forces on molecular motion in liquids. Picosecond rotational diffusion in alkanes and alcohols *J. Chem. Phys.* 87, 3739-48

Ben Amotz, D. & Drake, J.M., 1988, The solute size effect in rotational diffusion experiments: A test of microscopic friction theories *J. Chem. Phys.* 89, 1019-29

Benzler, J. & Luther, L., 1997, Rotational relaxation of biphenyl and *p*-terphenyl in *n*-alkanes: the breakdown of the hydrodynamic description *Chem. Phys. Lett.* 279, 333-38

Biasutti, M.A., Feyter, S.D., Backer, S.D., Dutt, G.B., Schryver, F.C.D., Ameloot, M., Schlichting P. & Müllen K., 1996, *Chem. Phys. Lett.* 248, 13-19

Birks, J.B., 1970, *Photophysics of Aromatic Molecules*, John Wiley & Sons, New York

Blanchard, G.J. & Wirth, M.J., 1986, Anomalous temperature-dependent reorientation of cresyl violet in 1-dodecanol J. Phys. Chem. *90, 2521-25*

Blanchard, G.J., 1987, Picosecond spectroscopic measurement of a solvent dependent change of rotational diffusion rotor shape *J. Chem. Phys.* 87, 6802-08

Blanchard, G.J. & Cihal, C.A. 1988, Orientational relaxation dynamics of oxazine 118 and resorufin in the butanols. Valence- and state-dependent solvation effects J. Phys. Chem. *92, 5950-54*

Blanchard, G.J. 1988, A study of the state-dependent reorientation dynamics of oxazine 725 in primary normal aliphatic alcohols J. Phys. Chem. *92, 6303-07*

Blanchard, G.J. 1989, State-dependent reorientation characteristics of Methylene Blue: the importance of dipolar solvent-solute interactions J. Phys. Chem. *93, 4315-19*

Blanchard, G.J. 1989, Detection of a transient solvent-solute complex using time-resolved pump-probe spectroscopy Anal. Chem. *61, 2394-98*

Bordewijk, P., 1980, *Z. Naturforsch.* 35A, 1207-17

Borin, I.A. & Skaf, M.S., 1998, *Chem. Phys. Lett.* 296, 125-30; 1999, Molecular association between water and dimethyl sulfoxide in solution: A molecular dynamics simulation study *J. Chem.Phys.* 110, 6412-20

Bossis, G., 1982, The dynamical Onsager cavity model and dielectric friction *Mol. Phys.* 46, 475-80

Bottcher, C.F.J. & Bordewijk, P., 1978, *Theory of Electric polarization*, 78th Edn., Vol. 11, Elsevier, Amsterdam

Bowman, R.M., Eisenthal, K.B. & Millar, D.P., 1988, Frictional effects on barrier crossing in solution: Comparison with the Kramers' equation *J. Chem. Phys.* 89, 762

Brito, P. & Bordewijk, P., 1980, Influence of dielectric friction on molecular reorientations *Mol. Phys.* 39, 217-226

Brocklehurst, B. & Young, R.N., 1995, Rotation of Perylene in Alkanes: Nonhydrodynamic Behavior J. Phys. Chem. 99, 40-43

Canonica, S., Schmid, A. & Wild, U.P., 1985, The rotational diffusion of *p*-terphenyl and *p*-quaterphenyl in non-polar solvents *Chem. Phys. Lett.* 122, 529-34

Chandler, D., 1974, Translational and rotational diffusion in liquids. I. Translational single-particle correlation functions *J. Chem. Phys.* 60, 3500-07; Translational and rotational diffusion in liquids. II. Orientational single-particle correlation functions *ibid* 3508-12

Chandra, A. & Bagchi, B., 1991, Molecular theory of solvation and solvation dynamics in a binary dipolar liquid *J. Chem. Phys.* 94, 8367-77

Chandra, A., 1995, Ion solvation dynamics in binary dipolar liquids: theoretical and simulation results for mixtures of Stockmayer liquids *Chem. Phys. Lett.* 235, 133-39

Chandrashekhar, K., Inamdar, S.R., Patil, D.C. & Math, N.N., 1993, Orientational relaxation of aminocoumarins by time-resolved dichroism with picosecond pulses *Spectrosc. Lett.* 28, 153-65

Chapman, C.F., Fee, R.S. & Maroncelli, M., 1990, Solvation dynamics in N-methylamides *J.Phys.Chem.* 94, 4929-35

Chen, S.H., Katsis, D., Schmid, A.W., Mastrangelo, J.C., Tsutsui, T. & Blanton, T.N., 1999, Circularly polarized light generated by photoexcitation of luminophores in glassy liquid-crystal films *Nature*, 397, 506

Chuang, J.T. & Eisenthal, K.B., 1971, Studies of effects of hydrogen bonding on orientational relaxation using picosecond light pulses *Chem. Phys. Lett.* 11, 368-70

Chuang, T.J. & Eisenthal, K.B., 1972, Theory of Fluorescence Depolarization by Anisotropic Rotational Diffusion *J. Chem. Phys.* 57, 5094-97

Cole, R.H., 1984, in *Molecular Liquids-Dynamics and Interactions*, Eds: A. J. Barnes, W. J. Orville-Thomas and J. Yarwood, pp. 59-100, Reidel, Dordrecht

Courtney, S.H., Kim, S.K., Canonica, S. & Fleming, G.R., 1986, Rotational diffusion of stilbene in alkane and alcohol solutions J. Chem. Soc. Faraday Trans. 82, 2065-72

Cowie, M.G. & Toporowski, P.M., 1961, Association in the binary liquid system dimethyl sulphoxide – water *Can. J. Chem.* 39, 2240-43

Dahm, W.J.A., Southerland, K.B. & Buch, K.A. 1991, Direct, high resolution, four-dimensional measurements of the fine scale structure of Sc≫1 molecular mixing in turbulent flows *Phys Fluids A*.3, 1115–1127

Day, T.J.F. & Patey, G.N., 1997, Ion solvation dynamics in binary mixtures *J. Chem. Phys.* 106, 2782-91

De la Torre, J.C., 1983, Biological actions and medical applications of dimethyl sulfoxide Ann. N.Y. Acad. Sci. 411, xi-xi

Debye, P., 1929, *Polar molecules*, Dover Publications, London

Demchenko, A.P., 2002, The red-edge effects: 30 years of exploration *Luminescence* 17, 19-42

Dote, J.L., Kivelson, D. & Schwartz, R.N., 1981, A molecular quasi-hydrodynamic free-space model for molecular rotational relaxation in liquids J. Phys. Chem. 85, 2169-80

Dutt, G.B., Doraiswamy, S., Periasamy, N. & Venkataraman, B., 1990, Rotational reorientation dynamics of polar dye molecular probes by picosecond laser spectroscopic technique *J. Chem. Phys.* 93, 8498-8513

Dutt, G.B. Konitsky, W. & Waldeck, D.H., 1995, Nonradiative relaxation of 2-phenylindene in solution and its implications for isomerization of stilbenes *Chem. Phys. Lett.* 245, 437-40

Dutt, G.B., Singh, M.K. & Sapre, A.V., 1998, Rotational dynamics of neutral red: Do ionic and neutral solutes experience the same friction? *J. Chem. Phys.* 109, 5994-5603

Dutt, G.B., Srivatsavoy, V.J.P. & Sapre, A.V., 1999, Rotational dynamics of pyrrolopyrrole derivatives in alcohols: Does solute–solvent hydrogen bonding really hinder molecular rotation? *J. Chem. Phys.* 110, 9623-29

Dutt, G.B. & Rama Krishna, G., 2000, Temperature-dependent rotational relaxation of nonpolar probes in mono and diols: Size effects versus hydrogen bonding *J. Chem. Phys.* 112, 4676-82

)utt, G.B. & Raman, S., 2001, Rotational dynamics of coumarins: An experimental test of dielectric friction theories *J. Chem. Phys.* 114, 6702-13

)utt, G.B. & Ghanty, T.K. 2003, Rotational Diffusion of Coumarins in Electrolyte Solutions: The Role of Ion Pairs *J. Phys. Chem B.* 107, 3257-64

)utt, G.B. & Ghanty, T.K., 2004, Is molecular rotation really influenced by subtle changes in molecular shape? *J. Chem. Phys.* 121, 3625-31

linstein, A., 1906, On the Theory of Brownian Motion (Zur Theorie der Brownschen Bewegung) *Ann. Phys.* 19, 371-81

lisenthal, K. B. 1975, Studies of chemical and physical processes with picosecond lasers *Acc. Chem. Res.* 8, 118-24

llsaesser, T. & Kaiser, W., 1991, Vibrational and Vibronic Relaxation of Large Polyatomic Molecules in Liquids *Annu. Rev. Phys. Chem. 42, 83-107*

lvans, G.T., Cole, R.G. & Hoffman, D.K., 1982, A kinetic theory calculation of the orientational correlation time of a rotorlike molecule in a dense fluid of spheres *J. Chem. Phys.* 77, 3209-20

lvans, G.T. & Evans, D.R., 1984, Kinetic theory of rotational relaxation in liquids: Smooth spherocylinder and rough sphere models *J. Chem. Phys.* 81, 6039-43

lvans, G.T., 1988, Translational and rotational dynamics of simple dense fluids *J.Chem. Phys.* 88, 5035-41

lee, R.S. & Maroncelli, M., 1994, Estimating the time-zero spectrum in time-resolved emmsion measurements of solvation dynamics *Chem. Phys.* 183, 235-47

lee, R.S., Milsom, J.A. & Maroncelli, M., 1991, Inhomogeneous decay kinetics and apparent solvent relaxation at low temperatures *J. Phys. Chem. 95, 5170-81*

lelderhof, B.U., 1983, Dielectric friction on a polar molecule rotating in a fluid *Mol. Phys.* 48, 1269-81; Dielectric friction on an ion rotating in a fluid *Mol. Phys.* 48, 1283-88

lleming, G.R., Morris, J.M. & Robinson, G.W., 1976, Direct observation of rotational diffusion by picosecond spectroscopy *Chem. Phys.* 17, 91-100

lleming, G.R., Knight, A.E.W., Morris, J.M., Robbins, R.J. & Robinson, G.W., 1977, Rotational diffusion of the mode-locking dye dodci and its photoisomer *Chem. Phys. Lett.* 49, 1-7

lleming, G.R., 1986, *Chemical Applications of Ultrafast Spectroscopy*, Oxford University Press: New York

lox, F. & Whittingham, K.P., 1974, Component interactions in aqueous dimethyl sulphoxide *J. Chem. Soc., Faraday Trans.*75, 1407-12

larg, S.K. & Smyth, C.P., 1965, Microwave Absorption and Molecular Structure in Liquids. LXII. The Three Dielectric Dispersion Regions of the Normal Primary Alcohols *J. Phys. Chem.* 69, 1294-1301

leirer, A. & Wirtz, K., 1953, Molecular theory of microfriction *Z. Naturforsch.* A8, 532-38

lordalla, B.C. & Zeidler, M.D., 1986; Molecular dynamics in the system water-dimethylsulphoxide *Mol. Phys.* 59, 817-28; 1991, NMR proton relaxation and chemical exchange in the system H16 2O/H17 2O-[2H6]dimethylsulphoxide, *Mol. Phys.* 74, 975-84

loulay, A.M., 1983, Rotational relaxation of OCS in n-alkanes: Collective and collisional effects *J. Chem. Phys.* 79, 1145-53

lustavsson, T., Cassara, L., Marguet, S., Gurzadyan, G., van der Meulen, P., Pommeret, S. & Mialocq, J.-C., 2003, *Photochem. Photobiol. Sci.* 2, 329

Hambir, S.A., Y. Jiang & Blanchard, G.J., 1993, Ultrafast stimulated emission spectroscopy of perylene in dilute solution: Measurement of ground state vibrational population relaxation *J. Chem. Phys.* 98, 6075-82

Hartman, R.S., Alavi, D.S. and Waldeck, D.H., 1991, An experimental test of dielectric friction models using the rotational diffusion of aminoanthraquinones J. Phys. Chem. *95, 7872-80*

Hartman, R.S., Konitsky, W.M., Waldeck, D.H., Chang, Y.J. & Castner, Jr, E.W., 1997, Probing solute–solvent electrostatic interactions: Rotational diffusion studies of 9,10-disubstituted anthracenes *J. Chem. Phys.* 106, 7920-30

Heilweil, E.J., Casassa, M.P., Cavanagh, R.R. & Stephenson, J.C., 1986, Population lifetimes of OH(v=1) and OD(v=1) stretching vibrations of alcohols and silanols in dilute solution *J. Chem. Phys.* 85, 5004-18

Heilweil, E.J., R. R. Cavanagh and J. C. Stephenson, 1987, Population relaxation of $CO(v = 1)$ vibrations in solution phase metal carbonyl complexes *Chem. Phys. Lett.* 134, 181-88

Heilweil, E.J., Casassa, M.P., Cavanagh, R.R. & Stephenson, J.C.1989, Picosecond Vibrational Energy Transfer Studies of Surface Adsorbates Annu. Rev. Phys. Chem. *40, 143-71*

Heilweil, E.J., Cavanagh, R.R. & Stephenson, J.C., 1989, CO(v=1) population lifetimes of metal–carbonyl cluster compounds in dilute CHCl$_3$ solution *J. Chem. Phys.* 89, 230-39

Horng, M.-L., Gardecki, J.A. & Maroncelli, M., 1997, Rotational Dynamics of Coumarin 153: Time-Dependent Friction, Dielectric Friction, and Other Nonhydrodynamic Effects J. Phys. Chem. A *101, 1030-47*

Hu, C.M. & Zwanzig, R., 1974, Rotational friction coefficients for spheroids with the slipping boundary condition *J. Chem. Phys.* 60, 4354-57

Hubbard, J.B. & Onsager, L., 1977, Dielectric dispersion and dielectric friction in electrolyte solutions. I. *J. Chem. Phys.* 67, 4850-57

Hubbard., J.B., 1978, Friction on a rotating dipole *J. Chem. Phys.* 69, 1007-09

Hubbard, J.B. & Wolynes, P.G., 1978, Dielectric friction and molecular reorientation *J. Chem. Phys.* 69, 998-1006

Huppert, D., Ittah, V. & E. Kosower, 1989, Static and dynamic electrolyte effects on excited-state behavior*Chem. Phys. Lett.* 159, 267-75

Huppert, D., Ittah, V. & Kosower, E., 1990, Static and dynamic electrolyte effects on excited large dipole solvation: high dielectric constant solvents *Chem. Phys. Lett.* 173, 496-502

Hynes, J.T., 1986, Chemical reaction rates and solvent friction *J. Stat. Phys.* 42, 149-168 and references therein

Imeshev, G. & Khundkar, L.R., 1995, Inhomogeneous rotational dynamics of a rodlike probe in 1-propanol *J. Chem. Phys.* 103, 8322-28

Inamdar, S.R., Chandrashekhar, K., Patil, D.C., Math, N.N. & Savadatti, M.I., 1995, Picosecond time-resolved laser emission of coumarin 102: Solvent relaxation, Pramana, J. Phys., 45, 279-290

Inamdar, S.R., Nadaf, Y.F. & Mulimani, B.G., 2003, Ground and excited state dipole moments of exalite 404 and exalite 417 uv laser dyes determined from solvatochromic shift of absorption and fluorescence spectra *J. Mol. Struct.* (*Theochem*), 624, 47-51

Inamdar, S.R., Mannekutla, J.R., Mulimani, B.G. & Savadatti, M.I., 2006, Rotational dynamics of nonpolar laser dyes *Chem. Phys. Lett.* 429, 141- 46

Inamdar, S.R., Gayathri, B.R. & Mannekutla, J.R., 2009, Rotational diffusion of coumarins in aqueous DMSO *J. Fluoresc.* 19, 693-703

ɔ, N., Kajimoto, O.& K. Hara, 2000, Picosecond time-resolved fluorescence depolarization of p-terphenyl at high pressures Chem. Phys. Lett. 318, 118-24

ɪrzeba, W., Walker, G.C., Johnson, A.E. & Barbara, P.F., 1991, Nonexponential solvation dynamics of simple liquids and mixtures Chem. Phys. 152, 57-68

ɑng, J. & Blanchard, G.J., 1994, Rotational Diffusion Dynamics of Perylene in n-Alkanes. Observation of a Solvent Length-Dependent Change of Boundary Condition J. Phys. Chem. 98, 6436-40; Vibrational Population Relaxation of Perylene in n-Alkanes. The Role of Solvent Local Structure in Long-Range Vibrational Energy Transfer, ibid, 9411-16: Vibrational Population Relaxation of Perylene in Its Ground and Excited Electronic States ibid, 9417-21

ɑng Y. & Blanchard, G.J., 1995, Vibrational Population and Orientational Relaxation Dynamics of 1-Methylpery lene in n-Alkanes. The Effective Range of Dipolar Energy Relaxation in Solution J. Phys. Chem. 99, 7904-12

ɑatze, K., Pottel, R. & Schaefer, M., 1989, Dielectric spectrum of dimethyl sulfoxide/water mixtures as a function of composition J. Phys. Chem. 93, 5623-27

ɑrasso, P.S. & Mungal, M.G., 1997, PLIF measurements in aqueous flows using the Nd:YAG laserExp Fluids 23, 382–387

ɑwski, A., Kuklinski, B. & Bojarski, P., 2005, Dipole moment of aniline in the excited S_1 state from thermochromic effect on electronic spectra Chem. Phys. Lett. 415, 251-55

ɑrov, A.S., Hurlbut, C., Dempsey, J.F., Shrinivas, S.B., Epstein, J.W., Binns, W.R., Dowkontt, P.F. & Williamson, J.F., 1999, Radiation Therapy Physics: Towards two-dimensional brachytherapy dosimetry using plastic scintillator: New highly efficient water equivalent plastic scintillator materials Med. Phys. 26, 1515-23

ɑoochesfahani, M.M. & Dimotakis, P.E., 1986, Mixing and chemical reactions in a turbulent liquid mixing layer J Fluid Mech. 170, 83–112

ɑrishnamurthy, M., Khan, K.K. & Doraiswamy, S., 1993, Rotational diffusion kinetics of polar solutes in hexamethylphosphoramide–water systems J. Chem. Phys. 98, 8640-47

ɑubinyi, M., Grofcsik, A., Kárpáti, T. & Jones, W.J., 2006, Rotational reorientation dynamics of ionic dye solutes in polar solvents with the application of a general model for the solvation shell Chem. Phys. 322, 247-54

ɑumar, P.V. & Maroncelli, M., 2000, The non-separability of "dielectric" and "mechanical" friction in molecular systems: A simulation study J. Chem. Phys. 112, 5370-81

ɑakowicz, J.R., 1983, Principles of Fluorescence Spectroscopy, Plenum Press, New York

ɑakowicz, J. R., 2006, Principles of fluorescence spectroscopy, Springer: New York

ɑaitinen, E., Korppi-Tommola, J. & Linnanto, J., 1997, Dielectric friction effects on rotational reorientation of three cyanine dyes in n-alcohol solutions J.Chem.Phys. 107, 7601-12

ɑaria, D. & Skaf, M., 1999, Solvation response of polar liquid mixtures: Water-dimethylsulfoxide J. Chem. Phys. 111, 300-09

ɑevitus, M, Negri, R.M. & Aramenda, P.F., 1995, Rotational Relaxation of Carbocyanines. Comparative Study with the Isomerization Dynamics J. Phys. Chem. 99, 14231-39

ɑingle Jr. R., Xu, X., Yu, S.C., Zhu, H. & Hopkins, J.B., 1990, Ultrafast investigation of condensed phase chemical reaction dynamics using transient vibrational spectroscopy: Geminate recombination, vibrational energy relaxation, and electronic decay of the iodine A' excited state J. Chem. Phys. 93, 5667-80

ɑuzar, A. & Chandler, D.,1993, Structure and hydrogen bond dynamics of water–dimethyl sulfoxide mixtures by computer simulations J. Chem. Phys., 98, 8160-73

Madden, P. & Kivelson, D., 1982, Dielectric friction and molecular orientation *J. Phys. Chem* 86, 4244-56

Mannekutla, J.R., Ramamurthy, P., Mulimani, B.G. & S.R. Inamdar, 2007, Rotationa dynamics of UVITEX-OB in alkanes, alcohols and binary mixtures *Chem. Phys.*, 340 149-57

Mannekutla, J.R., S. R. Inamdar, B. G. Mulimani and M. I. Savadatti, 2010, Rotationa diffusion of coumarins: A Dielectric friction study *J Fluoresc.* 20, 797–808

Maroncelli, M., 1993, The dynamics of solvation in polar liquids. *J. Molec. Liq.* 57, 1-37

Maroncelli, M. & Fleming, G.R., 1987, Picosecond solvation dynamics of coumarin 153: The importance of molecular aspects of solvation *J. Chem. Phys.* 86, 6221-39

Martin, D. & Hanthal, H., 1975, *Dimethyl Sulfoxide;* John Wiley & Sons, Inc., New York

McCarthy, P.K. & Blanchard, G.J., 1995, Vibrational Population Relaxation of Tetracene in n Alkanes. Evidence for Short-Range Molecular Alignment J. Phys. Chem. *99, 17748-53*

McCarthy, P.K. and Blanchard, G.J., 1996, Solvent Methyl Group Density Dependence o' Vibrational Population Relaxation in 1-Methylperylene: Evidence for Short-Range Organization in Branched Alkanes J. Phys. Chem. *100, 5182-87*

McMahon, D.R.A., 1980, Dielectric friction and polar molecule rotational relaxation *J. Chem Phys.* 72, 2411-24

Millar, D.P., Shah, R. & Zewail, A.H., 1979, Picosecond saturation spectroscopy of cresy violet: rotational diffusion by a "sticking" boundary condition in the liquid phase *Chem. Phys. Lett.* 66, 435-40

Moog, R.S., Ediger, M.D., Boxer, S. G. & Fayer, M.D., 1982, Viscosity dependence of the rotational reorientation of rhodamine B in mono- and polyalcohols. Picosecono transient grating experiments J. Phys. Chem. *86, 4694-4700*

Nadaf, Y.F., Mulimani, B.G., Gopal, M. & Inamdar, S.R., 2004, Ground and excited state dipole moments of some exalite dyes from solvatochromic method using solven polarity parameters *J. Mol. Struct. (Theochem)* 678, 177-81

Nee, T.W. & Zwanzig, R., 1970, Theory of Dielectric Relaxation in Polar Liquids *J. Chem Phys.* 52, 6353-63

Nowak, E., 1983, Dielectric friction and energy dissipation in polar fluids *J. Chem. Phys.* 79 976-81

Packer, K.J. & Tomlinson, D.J., 1971, Nuclear spin relaxation and self-diffusion in the binary system, dimethyl sulphoxide (DMSO)+ water J. Chem. Soc., Trans. Faraday 67 1302-14

Pal, A. & Daas, G., 2000, Excess molar volumes and viscosities of binary mixture: tetraethylene glycol dimethyl ether (tetraglyme) with chloroalkanes at 298.15K *J Mol. Liq.* 84, 327-37

Papazyan, A. & Maroncelli, M., 1995, Rotational dielectric friction and dipole solvation Tests of theory based on simulations of simple model solutions *J. Chem. Phys.* 102 2888-2919

Perrin, F., 1936, Mouvement brownien d'un ellipsoide (II). Rotation libre et depolarisatior des fluorescences. Translation et diffusion de moleculesellipsoidales, J. Phys Radium 7, 1-11

Phillips, L.A., Webb, S.P. & Clark, J.H., 1985, High-pressure studies of rotationa reorientation dynamics: The role of dielectric friction *J. Chem. Phys.* 83, 5810-21

Porter, G., Sadkowski, P.J. & Tredwell, C.J., 1977, Picosecond rotational diffusion in kinetic and steady state fluorescence spectroscopy *Chem. Phys. Lett.* 49, 416-20

Qunfang, L. & Yu-Chun, H., 1999, Correlation of viscosity of binary liquid mixtures. *Fluid Phase Equilibria* 154, 153–163 and references therein

Rice, S.A. & Kenney-Wallace, G.A., 1980, Time-resolved fluorescence depolarization studies of rotational relaxation in viscous media *Chem. Phys.* 47, 161-70

Rider, K.L. & Fixman, M., 1972, Angular Relaxation of the Symmetrical Top. II. The Rough Sphere *J. Chem. Phys.* 57, 2548-59

Roy, M. & Doraiswamy, S., 1993, Rotational dynamics of nonpolar solutes in different solvents: Comparative evaluation of the hydrodynamic and quasihydrodynamic models *J. Chem. Phys.* 98, 3213-23

Sanders, M.J. & Wirth, M.J., 1983, Evidence for solvation structural dependence of rotational diffusion anisotropy *Chem. Phys. Lett.* 101, 361-66

Safford, G.J., Schaffer, P.C., Leung, P.S., Doebbler, G.F., Brady G.W. & Lyden, E.F.X. 1969, Neutron Inelastic Scattering and X-Ray Studies of Aqueous Solutions of Dimethylsulphoxide and Dimethylsulphone *J. Chem. Phys.* 50, 2140-59

Shapiro, S.L. & Winn, K.R., 1980, Picosecond time-resolved spectral shifts in emission: dynamics of excited state interactions in coumarin 102 *Chem. Phys. Lett.* 71, 440-44

Sceats, M.G. & Dawes, J.M., 1985, On the viscoelastic properties of n-alkane liquids *J. Chem. Phys.* 83, 1298-1304

Selvaraju, C. & Ramamurthy, P., 2004, Excited-State Behavior and Photoionization of 1,8-Acridinedione Dyes in Micelles *Chem. Eur. J.* 10, 2253-62

Shank, C.V. & Ippen, E.P., 1975, Anisotropic absorption saturation with picosecond pulses *Appl. Phys. Lett.* 26, 62-63

Singh, M.K., 2000, Rotational Relaxation of Neutral Red in Alkanes: Effect of Solvent Size on Probe Rotation *Photochem. Photobiol.* 72, 438-43

Skaf, M. & Ladanyi, B.M., 1996, Molecular Dynamics Simulation of Solvation Dynamics in Methanol–Water Mixtures *J. Phys. Chem.* 100, 18258-68

Soper, A.K. & Luzar, A., 1996, Orientation of Water Molecules around Small Polar and Nonpolar Groups in Solution: A Neutron Diffraction and Computer Simulation Study *J. Phys. Chem.* 100, 1357-67

Soper, A.K. & Luzar, A., 1992, A neutron diffraction study of dimethyl sulphoxide–water mixtures *J. Chem.Phys.* 97, 1320-31

Spears, K.G. and L. E. Cramer, 1978, Rotational diffusion in aprotic and protic solvents *Chem. Phys.* 30, 1-8

Srivastava, A. & Doraiswamy, S., 1995, Rotational diffusion of rose bengal *J. Chem. Phys.* 103, 6197-6205

Steiner, R.F., 1991, *in Topics in Fluorescence Spectroscopy*, Vol. 2., J. R. Lakowicz (Ed.), Plenum Press, New York

Stokes, G., 1856, *Trans. Cambridge Philos. Soc.* 9, 5

Templeton, E. F. G., Quitevis, E. L. & Kenney-Wallace, G. A., 1985, Picosecond reorientational dynamics of resorufin: correlations of dynamics and liquid structure *J. Phys. Chem.* 89, 3238-43

Templeton, E.F.G. & Kenney-Wallace, G.A., 1986, Picosecond laser spectroscopic study of orientational dynamics of probe molecules in the dimethyl sulfoxide-water system *J. Phys. Chem.* 90, 2896-2900

Titulaer, U.M. & Deutch, J.M., 1974, Analysis of conflicting theories of dielectric relaxation *J. Chem. Phys.* 60, 1502-13

Tjai, T.H., Bordewijk P. & Bottcher, C.F.J.,1974, On the notion of dielectric friction in the theory of dielectric relaxation *Adv. Mol. Relax. Proc.* 6, 19-28

Tokuhiro, T., Menafra L. & Szmant, H.H.,1974, Contribution of relaxation and chemical shift results to the elucidation of the structure of the water-DMSO liquid system *J. Chem Phys*.61, 2275-82

Traube, J., 1886, Ber, Dtsch Chem.Ges. B.19, 871-892

Vaisman, I.I. & Berkowitz, M.L., 1992, Local structural order and molecular associations in water-DMSO mixtures. Molecular dynamics study *J. Am. Chem. Soc.* 114, 7889-96

van der Zwan, G. & Hynes, J.T., 1985, Time-dependent fluorescence solvent shifts, dielectric friction, and nonequilibrium solvation in polar solvents *J. Phys. Chem.* 90, 4181-88

Valenta, J., Dian, J., Hála, J., Gilliot, P. &Lévy, R., 1999 Persistent spectral hole-burning and hole-filling in CuBr semiconductor nanocrystals *J. Chem. Phys.* **111**, 9398-403

Voigt, W., 2005, Sulforhodamine B assay and chemosensitivity *Methods Mol. Med.* 110, 39-48

von Jena, A. & Lessing, H. E., 1979a, Rotational-diffusion anomalies in dye solutions from transient-dichroism experiments *Chem. Phys.* 40, 245-56

von Jena, A. & Lessing, H. E., 1979b, Rotational Diffusion of Prolate and Oblate Molecules from Absorption Relaxation *Ber. Bunsen-Ges. Phys. Chem.* 83, 181-91

von Jena, A. & Lessing, H.E., 1981, Rotational diffusion of dyes in solvents of low viscosity from transient-dichroism experiments *Chem. Phys. Lett.* 78, 187-93

Wagener, A. & Richert, R., 1991, Solvation dynamics versus inhomogeneity of decay rates as the origin of spectral shifts in supercooled liquids *Chem. Phys. Lett.* 176, 329-34

Waldeck, D.H. and G. R. Fleming, 1981, Influence of viscosity and temperature on rotational reorientation. Anisotropic absorption studies of 3,3'-diethyloxadicarbocyanine iodide *J. Phys. Chem.* 85, 2614-17

Waldeck, D.H., Lotshaw, W.T., McDonald, D.B. & Fleming, G.R., 1982, Ultraviolet picosecond pump-probe spectroscopy with a synchronously pumped dye laser. Rotational diffusion of diphenyl butadiene *Chem. Phys. Lett.* 88, 297-300

Widom, B., 1960, Rotational Relaxation of Rough Spheres, *J. Chem. Phys.* 32, 913-23

Wiemers, K. & Kauffman, J. F., 2000, Dielectric Friction and Rotational Diffusion of Hydrogen Bonding Solutes *J. Phys. Chem. A* 104, 451-57

Williams, A.M., Jiang, Y. & Ben-Amotz, D., 1994, Molecular reorientation dynamics and microscopic friction in liquids *Chem. Phys.* 180, 119-29

Yip, R.W., Wen, Y. X. & Szabo, A.G., 1993, Decay associated fluorescence spectra of coumarin 1 and coumarin 102: evidence for a two-state solvation kinetics in organic solvents *J. Phys. Chem.* 97, 10458-62

Yoshimori, A., Day, T.J.F. & Patey, G.N., Theory of ion solvation dynamics in mixed dipolar solvents *J. Chem. Phys.* 109, 3222-31

Youngren, G.K. & Acrivos, A., 1975, Rotational friction coefficients for ellipsoids and chemical molecules with the slip boundary condition *J. Chem. Phys.* 63, 3846-48

Zwanzig, R. & Harrison, A.K., 1985, Modifications of the Stokes–Einstein formula *J. Chem. Phys.* 83, 5861- 62

8

Flow Evolution Mechanisms of Lid-Driven Cavities

José Rafael Toro and Sergio Pedraza R.
Grupo de Mecánica Computacional, Universidad de Los Andes
Colombia

Introduction

The flow in cavities studies the dynamics of motion of a viscous fluid confined within a cavity in which the lower wall has a horizontal motion at constant speed. There exist two important reasons which motivate the study of cavity flows. First is the use of this particular geometry as a benchmark to verify the formulation and implementation of numerical methods and second the study of the dynamics of the flow inside the cavity which become very particular as the Reynolds (Re) number is increased, i.e. decreasing the fluid viscosity.

Most of the studies, concerning flow dynamics inside the cavity, focus their efforts on the steady state, but very few study the mechanisms of evolution or transients until the steady state is achieved (Gustafson, 1991). Own to the latter aproach it was considered interesting to understand the mechanisms associated with the flow evolution until the steady state is reached and the steady state per se, since for different Re numbers (1,000 and 10,000) steady states are "similar" but the transients to reach them are completely different.

In order to study the flow dynamics and the evolution mechanisms to steady state the Lattice Boltzmann Method (LBM) was chosen to solve the dynamic system. The LBM was created in the late 90's as a derivation of the Lattice Gas Automata (LGA). The idea that governs the method is to build simple mesoscale kinetic models that replicate macroscopic physics and after recovering the macro-level (continuum) it obeys the equations that governs it i.e. the Navier Stokes (NS) equations. The motivation for using LBM lies in a computational reason: Is easier to simulate fluid dynamics through a microscopic approach, more general than the continuum approach (Texeira, 1998) and the computational cost is lower than other NS equations solvers. Also is worth to mention that the prime characteristic of the present study and the method itself was that the primitive variables were the vorticity-stream function not as the usual pressure-velocity variables. It was intended, by chosing this approach, to understand in a better way the fluid dynamics because what characterizes the cavity flow is the lower wall movement which creates itself an impulse of vorticiy which is transported within the cavity by diffusion and advection. This transport and the vorticity itself create the different vortex within the cavity and are responsible for its interaction.

In the next sections steady states, periodic flows and feeding mechanisms for different Re numbers are going to be studied within square and deep cavities.

2. Computational domains

The flow within a cavity of height **h** and wide **w** where the bottom wall is moving at constan velocity U_0 Fig.1 is going to be model. The cavity is completely filled by an incompresibl fluid with constant density ρ and cinematic viscosity ν.

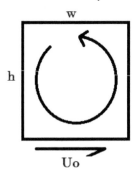

Fig. 1. Cavity

3. Flow modelling by LBM with vorticity stream-function variables

Is important to introduce the equations that govern the vorticity transport and a few definitions that will be used during the present study.

Definition 0.1. *A vortex is a set of fluid particles that moves around a common center*

The vorticity vector is defined as $\omega = \nabla \times v$ and its transport equation is given by

$$\frac{\partial \omega}{\partial t} + [\nabla \omega]v = [\nabla v]\omega + \nu\nabla^2\omega. \tag{1}$$

which is obtained by calculating the *curl* of the NS equation. For a 2D flow Eq.(1) is simplifiec to obtain

$$\frac{\partial \omega}{\partial t} + [\nabla \omega]v = \nu\nabla^2\omega. \tag{2}$$

In order to recover the velocity field from the vorticity field the Poisson equation for the stream function needs to be solved. The Poisson equation wich involves the stream function is statec as

$$\nabla^2\psi = -\omega \tag{3}$$

where ψ is the stream function who carries the velocity field information as

$$u = \frac{\partial \psi}{\partial y} \ , \ v = -\frac{\partial \psi}{\partial x} \ . \tag{4}$$

and ensures the mass conservation. The motivation for adopting vorticity as the primitive variables lies in the fact that every potential, as the pressure, is eliminated which is physicaly desirable because being the vorticity an angular velocity, the pressure, which is always norma to the fluid can not affect the angular momentum of a fluid element.

3.1 Numerical method

Consider a set of particles that moves in a bidimensional lattice and each particle with a finite number of movements. Now a vorticity distribution function $g_i(x, t)$ will be asigned to each particle with unitary velocity e_i giving to it a dynamic consistent with two principles:

1. Vorticity transport
2. Vorticity variation in a node own to particle collision

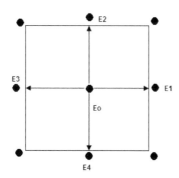

Fig. 2. D2Q5 Model.2 dimensions and 5 possible directions of moving

Observation 0.2. *The method only considers binary particle collisions.*

The evolution equation is discribed by

$$g_k(\vec{x} + c\vec{e}_k \Delta t, t + \Delta t) - g_k(\vec{x}, t) = -\frac{1}{\tau}[g_k(\vec{x}, t) - g_k^{eq}(\vec{x}, t)]^1 \tag{5}$$

where e_k are the possible directions where the vorticity can be transported as shown in Fig.2. $c = \Delta x / \Delta t$ is the fluid particle speed, Δx and Δt the lattice grid spacing and the time step respectively and τ the dimensionless relaxation time. Clearly Eq.(5) is divided in two parts, the first one emulates the advective term of (1) and the collision term, which is in square brackets, emulates the diffusive term of equation (1).
The equilibrium function is calculed by

$$g_k^{eq} = \frac{w}{5}[1 + 2.5\frac{\vec{e}_k \cdot \vec{u}}{c}]. \tag{6}$$

The vorticity is calculed as

$$w = \sum_{k \geq 0} g_k \tag{7}$$

and τ, the dimensionless relaxation time, is determined by Re number

$$Re = \frac{5}{2c^2(\tau - 0.5)}. \tag{8}$$

---

[1] The evolution equations were taken from (Chen et al., 2008) and (Chen, 2009). Is strongly recomended to consult the latter references for a deeper understanding of the evolution equations and parameter calculations.

In order to calculate the velocity field Poisson equation must be solved (3). In order to do this (Chen et al., 2008) introduces another evolution equation.

$$f_k(\vec{x} + c\vec{e}_k \Delta t, t + \Delta t) - f_k(\vec{x}, t) = \Omega_k + \hat{\Omega}_k. \tag{9}$$

Where

$$\Omega_k = \frac{-1}{\tau_\psi}[f_k(\vec{x}, t) - f_k^e q(\vec{x}, t)], \quad \hat{\Omega}_k = \Delta t \zeta_k \theta D \tag{10}$$

and $D = \frac{c^2}{2}(0.5 - \tau_\psi)$. τ_ψ is the dimensionless relaxation time of the latter evolution equation wich can be chosen arbitrarly. For the sake of understanding the evolution equations, the equation (9) consist on calculating $\frac{D\psi}{Dt} = \nabla^2 \psi + \omega$ until $\frac{D\psi}{Dt} = 0$, having found a solution ψ for the Poisson equation.

By last, the equlibrium distribution function is defined as

$$f_k^{eq} = \begin{cases} \zeta_k \psi & k = 1, 2, 3, 4 \\ -\psi & k = 0 \end{cases} \tag{11}$$

where ξ_k and ζ_k are weight parameters of the equation.

3.2 Algorithm implementation
In order to implement the evolution equation Eq.(5) two main calculations are considered. First, the collision term is calculated as

$$g_k^{int} = -\frac{1}{\tau}[g_k(\vec{x}, t) - g_k^{eq}(\vec{x}, t)] \tag{12}$$

and next the vorticity distributions is transported as

$$g_k(\vec{x} + c\vec{e}_k \Delta t, t + \Delta t) = g_k^{int} + g_k(\vec{x}, t) \tag{13}$$

which is, as mentioned, the basic concept that governs the LBM, collisions and transportation of determined distribution in our case a vorticity distibution.

3.2.1 Algorithm and boundary conditions
1. Paramater Inicialization
   - Moving wall velocity: $U_0 = 1$.
   - $\psi|_{\partial\Omega} = 0$, own to the fact that no particle is crossing the walls.
   - $u = v = 0$ in the whole cavity excepting the moving wall.
   - Re number definition[2]
2. Wall vorticity calculation

$$\omega|_{\partial\Omega} = \frac{7\psi_w - 8\psi_{w-1} + \psi_{w-2}}{2\Delta n^2} \tag{14}$$

$$\omega|_{\partial\Omega} = \frac{7\psi_w - 8\psi_{w-1} + \psi_{w-2}}{2\Delta n^2} - \frac{3U_0}{\Delta n} \tag{15}$$

Both equations came from solving Poisson equation Eq.(3) on the walls by a second order Taylor approximation. Eq.(15) is used on the moving wall nodes.

---

[2] For the sake of clarity Re number is imposed in the method by the user which intrinsically is imposing different flow viscosities.

3. Velocity field calculation using Eq.(4)
4. Equilibrium probability calculation using Eq.(6)
5. Colission term calculation using Eq.(12)
6. Probability transport using Eq.(13)
7. Vorticity field calculation using Eq.(7)
8. Solution of Poisson equation: In order to solve Poisson equation the evolution equation Eq.(9) for the stream-function distribution was implemented within a loop wishing to compare f_k's values (i.e. ψ) aiming to achive that $\frac{D\psi}{Dt} = \nabla^2\psi + \omega = 0$. For the latter loop the process terminated when

$$\sum_{x,y} |f_k^+ - f_k| < 10^{-3}.$$

While the simulations were ran, it was found that the algorithm was demanding finer meshes for higher Re numbers, i.e. 700x700 nodes mesh for Re 6,000, increasing the computational cost and most of the times ending in overflows own to the fluid regime. To overcome this situations a turbulence model was introduced to the LBM proposed by (Chen, 2009).

4. Introduction of turbulence in LBM

The principal characteristic of a turbulent flow is that its velocity field is of random nature. Considering this, the velocity field can be split in a deterministic term and in a random term i.e. $U(x,t) = \bar{U}(x,t) + u(x,t)$, being the deterministic and random term respectively. In order to solve the velocity field, the NS equations are recalculated in deterministic variables adding to the set a closure equation own to the loss of information undertaken by solving only the deterministic term. At introducing a turbulent model there exist three different approaches: algebraic models, closure models and Large Eddy Simulations (LES) being the latter used in the present study. LES were introduce by James Deardorff on 1960 (Durbin & Petersson-Rief, 2010). Such simulations are based in the fact that the bulk of the system energy is contained in the large eddys of the flow making not neccesary to calculate all the vortex disipative range which would imply a high computational cost (Durbin & Petersson-Rief, 2010). If small scales are ommited, for example by increasing the spacing by a factor of 5, the number of grid points is substantially reduced by a factor of 125 (Durbin & Petersson-Rief, 2010). In LES context the elimination of these small scales is called filtering. But this filtering or omission of small scales is determined as follows: the dissipative phenomenon is replaced by an alternative that produces correct dissipation levels without requiring small scale simulations. The Smagorinsky model was introduced where another flow viscosity (usually known as subgrid viscosity) is considered which is calculated based on the fluid deformation stress. Specifically it is model as $\nu_t = (C\Delta)^2|S|$ Chen et al. (2008) where

$$S_{ij} = \frac{1}{2}\left(\frac{\partial \bar{U}_i}{\partial x_i} + \frac{\partial \bar{U}_j}{\partial x_j}\right),$$

Δ is the filter width and C the Smagorinsky constant. In the present study $C = 0.1$ and $\Delta = \Delta x$. Assuming this new **subgrid viscosity** ν_t the momentum equation is given by

$$\frac{\partial \omega}{\partial t} + [\nabla \omega]v = \frac{\partial}{\partial x}\left(\nu_e \frac{\partial \omega}{\partial x}\right) + \frac{\partial}{\partial y}\left(\nu_e \frac{\partial \omega}{\partial y}\right)$$

where

$$v_e = v_t + v.$$

As the transport equation has changed, the LBM evolution equation has also changed

$$g_k(\vec{x} + c\vec{e}_k\Delta t, t + \Delta t) - g_k(\vec{x}, t) = -\frac{1}{\tau_e}[g_k(\vec{x}, t) - g_k^{eq}(\vec{x}, t)] \tag{16}$$

where

$$\tau_e = \tau + \frac{5(C\Delta)^2|S|}{2c^2\Delta t} \quad and \quad |S| = |\omega|^3.$$

Having a new evolution equation Eq.(16) the algorithm has to be modified adding a new step where τ_e is calculated based on the vorticity field. After making this improvement to the method, the algorithm began to work eficiently allowing to achive higher Re numbers without compromising the computer cost, justifing the use of a LBM.

5. Steady state study for different Re numbers

It is said that the flow has reached steady state when collisions and transport do not affect each node probability. Concerning the algorithm it was considered that the flow had reached the steady state when its energy had stabilized and when the maps of vorticity and stream function showed no changes through time.

Steady state vortex configuration for Re 1,000 and Re 10,000 is shown in Fig.3. It worth to notice that both are very similar, a positive vortex that fills the cavity and two negative vortices at the corners of the cavity. This configuration was observed from Re 1,000 to Re 10,000 being a prime characteristic of cavity flows. It is also important to clarify that for Re 10,000 the steady state presents a periodicity which is located in the upper left vortex that we shall see later, indeed Fig.3(b) is a "snapshot" of the flow.

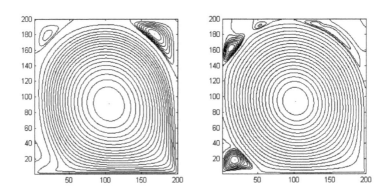

(a) Stream-function map in steady state for Re 1,000. (b) Stream-function map in steady state for Re 10,000.

Fig. 3. Steady states. Maps were taken at 100,000 and 110,000 iterations respectively.

---

[3] Is strongly recomended to consult (Chen, 2009) for a deeper understanding of the evolution equations and parameter calculations.

.1 Deep cavities

everal studies have proposed to study the deep cavity geometry (Gustafson, 1991; Patil et al., 006) but none has reached to simulate high Re numbers possibly because the mesh sizes. Due o the LBM low computational cost it was decided to present the study of a deep cavity with n aspect ratio (AR) of 1.5 for Re 8,000.

.1.1 Vortex dynamics

 general description is presented emphasizing the most important configurations through volution to steady state:

Step 1 Fig.4(a) The positive vortex creates a negative vortex that arises from the right wall triggering an interaction since the begining of the evolution.

Step 2 Fig.4 (b) The negative vortex that arises from the right wall has taken the whole cavity confining the positive vortex to the bottom.

Step 3 Fig.4(c,d) Positive vortices have joined in one by an interesting process discribed in Sec6. This union creates a "mirror" phenomenon inside the cavity.

Step 4 Fig.4(e) The positive vortex expands into the cavity moving upward the negative vortex until the steady state is reached in which both vortices occupy the same space of the cavity. Is worth to notice that this vortex distribution is not achieved in the square cavity steady state.

.1.2 Mirror phenomenon

During the evolution it was observed that after positive vortices joined (Fig.4(c, d)) the new ig positive vortex acted as a moving wall for the negative vortex injecting vorticity to it. Reproducing the behavior seen in the square cavity, now by the negatie vortex. Ergo a *quasi quare cavity* was created in the top of the cavity but instead having a moving wall it had a ortex. The phenomenon is shown in Fig.5 where it is clear that the top of the deep cavity is "reflection" of the square cavity with respect to an imaginary vertical axis drawn between hese two.

. Vortex binding

 particular process for Re 10,000 in square and deep cavities was found to take place through volution. This process occurs several times throughout evolution, named Vortex Binding. n this process isolated vortices get connected forming a "massive" vortex which eventually vill configure the steady state vortices distribution. A binding process that occured through volution is shown in Fig.6 binding a positive vortex that appeared in the upper right corner vith the positive vortex that came from the movement of the bottom wall. n order to explain the binding process, which is illustrated in Fig.6, recall the vorticity ransport equation Eq.(1). The transport equation is divided in two terms that dictate the ransport of vorticity, the diffusive term $\nu\nabla^2\omega$ and the advective term $[\nabla\omega]v$. For a high Re umber flow the diffusive term can be neglected, turning the attention in the advective term. As the flow evolved it was seen that the vorticity and stream-function contour lines tended to lign as shown in Fig.7(a) making the vorticity gradient vector and velocity vector orthogonal t different places (Fig.7(a)) causing $[\nabla\omega]v = 0$, i.e. no vorticity transport. As shown in Fig.7(b) vorticity contour lines started to curve, due to its own vorticity, crossing vith the stream-function contour lines and making $[\nabla\omega]v \neq 0$. In Fig.7(b)can be seen that

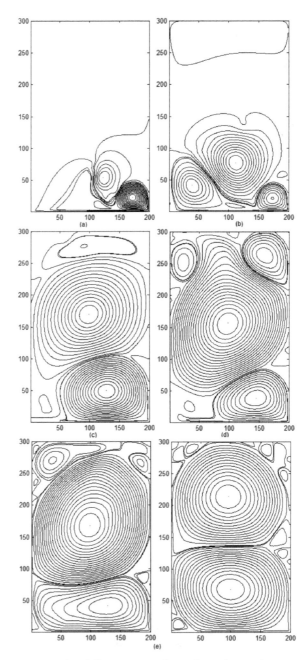

Fig. 4. Stream-function map for different times through evolution for a cavity with AR=1.5
and Re 8,000 in a 200x300 nodes mesh. *a,b,c,d and e* were taken at 20,000, 50,000, 150,000,
180,000 and 260,000-340,000 iterations.

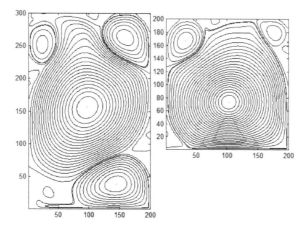

Fig. 5. *Left* Stream-function map for Re 8,000 in a cavity with AR=1.5 (200x300 nodes) *Right* Stream-function map in a square cavity for Re 8,000 (200x200 nodes).

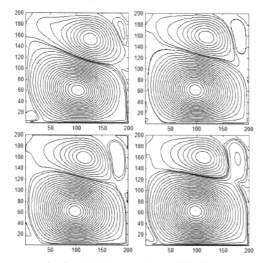

Fig. 6. Stream-function maps for Re 10,000 were Vortex binding process take place. Four maps were taken between 80,000 and 90,000 iterations

the vorticity gradient and the velocity vector are no longer orthogonals creating vorticity transport in different places which made possible the vortex binding to take place.

7. Periodicity in cavity flows

In the study of dynamic systems, being the case of the present study the NS equations, and their solutions there exist bifurcations leading to periodic solutions. Specifically in cavity flows, when the Re number is increased, such bifurcations take place known as **Hopf Bifurcations**. Willing to understand how this Bifurcation takes place the *Sommerfelds* infinitesimal perturbation model is introduced. This perturbation model considers a small

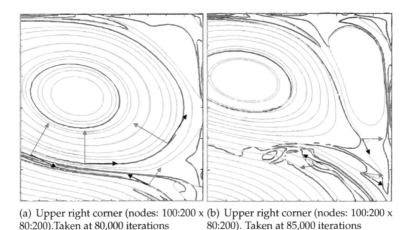

(a) Upper right corner (nodes: 100:200 x (b) Upper right corner (nodes: 100:200 x
80:200).Taken at 80,000 iterations   80:200). Taken at 85,000 iterations

Fig. 7. *Left* Stream-function contour lines (**Green**), vorticity contour lines(**Red**), vorticity
gradient(**Red**), velocity vector(**Black**).*Right* Stream-function contour lines (**Green**), vorticity
contour lines(**Red**), vorticity gradient(**Red**), velocity vector(**Black**)and angle between $[\nabla \omega]$
and v(**Blue**)

perturbation of the dynamical system in order to study the equilibrium state or the lack of it.
Let be considered the next dynamical system

$$\frac{d\acute{u}}{dt} = [M_\nu]\,\acute{u}. \tag{17}$$

The solution of Eq.(17) lies on finding the eigenvectors of the $[M_\nu]$ operator which is in
function of the fluid vicosity. Depending on the Re number the eigenvalues (and eigenvector)
can be complex i.e. $\lambda \in \mathbb{C}$, leading to periodic solutions(Toro, 2006) or Bifurcations. In
(Auteri et al., 2002) the bifurcation for a cavity flow was located between 8017,6 and 8018,8
(Re numbers) but since 1995 (Goyon, 1995) reported the existence of particular periodic flow
located in the upper left corner of a square cavity. In order to find the flow periodicity for Re
10,000 and determine if the system had reached its asymptotic state the system energy was
used as a measure. A Periodic flow for a deep cavity is shown in Fig.8[4]

8. Flow transients

Studying vorticity and stream-function maps was found that the way to get to the *same* state
in most of the flow (Fig.3(a) and Fig.3(b)), with the exception of the corners for Re 10,000
which oscillate, change significantly as the number of Re varies. In order to illustrate this
"bifurcation" vorticity transients for Re 1,000 and Re 10,000 are shown in Figs.9, 10 and 11
until steady state configuration is reached.

8.1 Transient description
For Re 1,000 the positive vortex is created on the lower right corner by the bottom wall
movement. Latter vortex is feed and grows until the whole cavity is taken cornering

---

[4] A well discribed periodic flow for square cavity can be found in (Goyon, 1995).

nd breaking a negative vortex that has accompanied it since the beginning of evolution vithout qualitative form changes, only scaling the first configuration until the steady state onfiguration is achieved in Fig.3(a).

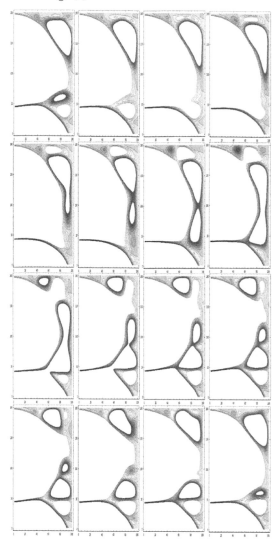

ig. 8. Stream-function maps for a deep cavity with AR=1.5 and Re 8,000 where periodic flow ake place. Maps were taken between 300,000 and 309,000 iterations. *White patches are vortices with high absolute vorticity.* Cavity upper right corner (100:200x100:300) nodes, see ig.4(e-right)

or Re 10,000 the positive vortex is created due to the lower wall movement and immediately tself creates a negative vortex coming from the right wall. Unlike Re 1,000 these two

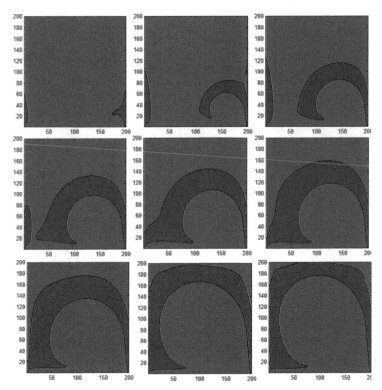

Fig. 9. Vorticity maps: Positive vorticiy (**Blue**), Negative vorticity(**Red**) (200x200 nodes mesh). The nine maps were taken at 10,000, 20,000, 30,000, 40,000, 50,000, 60,000, 70,000, 80,000, 100,000 and 110,000 iterations respectively.

vortices qualitatively change during evolution, changing size and shape until the stable state configuration is reached shown in Fig.3(b).

It is worth to notice that vorticity maps for Re 1,000 and Re 10,000 are topologicaly very different. For Re 1,000 no interaction between positive and negative vorticity is presented but for Re 10,000 interaction is presented since the begining of evolution until the steady state and in the steady state itself because what causes the flow periodicity is the interaction of positive and negative vortices on the corners of the cavity.

9. Vortex feeding mechanisms

Cavity flow is a phenomenon characterized by a continuos vorticity injection to the system induced by the moving wall. The vorticity arises because the no-slip condition (viscous fluid) creating an impulse of vorticity that is transported into the cavity by advection or diffusion Eq.(1). As seen since the beginning the vorticity transport equation is divided in a diffusive term $\nu\nabla^2\omega \approx \frac{1}{Re}\nabla^2\omega$ and in an advective term $[\nabla\omega]\,v$. At the beginning of the flow evolution the vorticity input is transported from the wall purely by diffusion but as the flow evolves both terms of the vorticity transport equation start to have different weights, being the diffusive term the most sensitive to Re number variations.

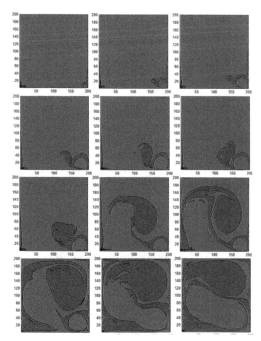

Fig. 10. Vorticity maps: Positive vorticiy (**Blue**), Negative vorticity(**Red**) (200x200 nodes mesh). The twelve maps were taken from 10,000 to 60,000 iterations.

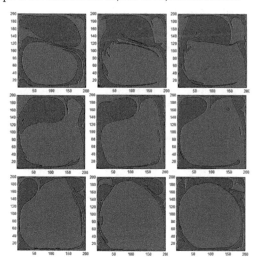

Fig. 11. Vorticity maps: Positive vorticiy (**Blue**), Negative vorticity(**Red**) (200x200 nodes mesh). The nine maps were taken from 60,000 to 110,000 iterations.

Definition 0.3. *A **vorticity channel** is a bondary layer, coming from a wall, that feeds and creates vortex.*

9.1 Channel creation and some other characteristics

Channel creation is derived from two different phenomena: First is the energy transformation that occurs in the wall because the system continually transforms translational energy into rotational energy. Secondly a vortex whatever its sign is creates a channel of opposite sign. In oder to understand the latter suppose a positive vortex near a wall. The vortex make the particles that lie between it and the wall start spinning or *rotate*, due to viscosity, in the opposite direction causing a vorticity input - in this case negative - to the system.

There are three important features on the channels. The first and most important is that the channels transport vorticity from the walls inside the cavity and also diffuses vorticity along the route to nearby channels in proportion to the existing vorticity gradient. Secondly a positive channel always wraps a negative vortex and a negative channel always wraps a positive vortex. And finally channel thickness is function of the Re number.

9.2 Channel study for Re 1,000

- **Channel creation:** The transient is shown in Fig.9. Since the beginning there is a feeding channel from the right wall that grows merging in a left wall channel. It is worth noticing that the channel wraps the positive vortex during evolution (Fig.12(a)) but never interacts with it.

- **Channel characteristics:** In Fig.9 can be observed that the feeding channels are thick. This ows to the fact the diffusive term of the transport equation is big enough to let vorticity be spread within the fluid apart from being transported.

9.3 Channel study for Re 10,000

Before studying the channels it is worth to clarify that in Fig.10 and 11 channels are the thin red "tubes" and the color patches are formed vortices which are fed by channels.

- **Channel creation:** In the transient shown in Fig.10 can be seen since the beginning the appearance of a feeding channel coming from the right wall, but unlike the Re 1,000 transient, it begins to feed a vortex (sixth square of transient Fig.10) that grows inside the cavity. This vortex has the ability to interact in different ways (Fig.10 and 11) with the positive vortex that eventually will take the cavity. What is interesting about the vortex interaction, apart from the different forms that arise in the transient, is that the latter vortex has as many vorticity as the positive one, allowing them to interact in many ways. This interaction is able to produce a configuration seen in the deep cavity steady state where both vortices occupy the cavity without cornering each other but highly unstable (twelfth square Fig.10). This occurs because the diffusive term of the transport equation has less weigth, allowing to concentrate vorticity without being spread across the cavity, which is the case for Re 1,000. It is also important to mention that for Re 10,000 negative channel wraped positive vortex and vice versa (Fig.12(b)) as happens for Re 1,000.

- **Channel characteristics:** Unlike Re 1,000 channels the thickness of Re 10,000 channels are smaller, due to the diffusive low weight term in the vorticity transport equation.

10. Circulation study for different Re numbers

In order to understand more about what is happening with the vorticity of the system was decided to study the circulation behavior. The circulation is defined as $\Gamma = \int \omega dA$. An interesting aspect of the circulation is that, although it must be constant in the system over

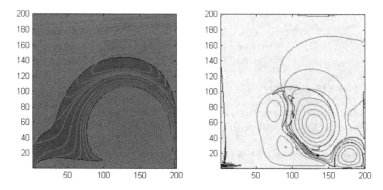

(a) Superposition for Re 1,000 during (b) Superposition for Re 10,000 during
evolution.   evolution

Fig. 12. Stream-function contour lines (blue) and vorticity maps superposition. *Left* Positive
vorticity (Dark red) Negative vorticity (Light red), *right* Positive vorticity (Aqua) Negative
vorticity (Aquamarine).

time according to Kelvins theorem, it can be split into positive and negative values. As seen,
the prime characteristic of the flow is the positive vorticity input from the lower wall deriving
n positve circulation diferential.

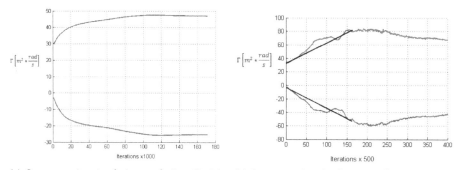

(a) Square cavity circulation evolution. Positive (b) Square cavity circulation evolution. Positive
Γ (Red) and negative Γ (Blue) Γ (Red) and negative Γ (Blue)

Fig. 13. *Left* Square cavity circulation for Re 1,000. *Right* Square cavity circulation for Re
10,000.

In both figures can be seen that the flow reaches a maximum around the 100.000 iterations
when the positive vortex has taken all the cavity (Fig.3.1 and 3.2). What is interesting are the
values of circulation that are achieved for each value of Re (Table.1).
Several important things are shown in Table.1. First the circulation increase for Re 10,000 is
three times bigger than Re 1,000 i.e. $\Delta\Gamma_{Re1,000} = 18.36$ compared with $\Delta\Gamma_{Re10,000} = 50.5$.
Latter observation means that as the viscosity decreases the system is able to accumulate
more circulation. Finally, system circulation is consistent whit Kelvin's theorem even though

	Re 1,000		Re 10,000	
	max	min	max	min
Positive Γ	48.52	30.16	83.5	33
Negative Γ	23.8	3.09	60.67	2.55

Table 1. Circulation values comparison

positive circulation increases negative circulation increases too maintaining a circulation differential of about 30 throughout evolution (Fig.13 a and b).

10.1 Why does the circulation fall after rising for Re 10,000?

It can be seen in Fig.13 that for Re 1,000 positive (negative) circulation reaches its maximum (minimum) and stabilizes around latter value, which fails to happen for Re 10,000 where circulation peaks at a "constant" rate but after reaching maximum starts decreasing. The motivation of this subsection is to explain why this change of slope took place (Fig.13(b)) and try to predict it analiticaly because it was observed that for different Re numbers the same change in slope occures reaching different values of maximum circulation.

In order to understand this phenomena recall that the cavity has vorticity channels that feed and remove vorticity into and out the system affecting the circulation values. Having mentioned this observation and due to the low weight diffusive term has in the transport equation, $\frac{d\Gamma}{dt}$ is calculed according to the gradient of vorticity on the walls (18), which is the same as quantifying how much vorticity is entering and leaving the system.

$$\frac{d\Gamma}{dt} = \int_{\partial\Omega} \nabla w \cdot n ds \tag{18}$$

After ploting Eq.(18) through time it was found that $\frac{d\Gamma}{dt}$ was constant until 100.000 iterations, which is when the positive vortex has taken the cavity, reflecting the "constant" increase of circulation Fig.13(b). More interesting and contradicting the assumption made was that $\frac{d\Gamma}{dt}$ does not fall after the 100,000 iterations, situation that was expected since a slope change was observed in the Fig.13(b) after 100,000 iterations. Willing to explain this behavior the following hypothesis was proposed:

Assume a unit of vorticity entering to the system Fig.14.

This unit feeds the positive vortex. The vortex is not able to accumulate more circulation, as it has reached the steady state configuration therefore this unit of vorticty has to be "passed" to each of the corner vortices, which also are not able to accumulate more circulation having to pass it to the upper wall and balancing the accounts of vorticity on the walls. Since the way of calculating the $\frac{d\Gamma}{dt}$ is based on counting how much vorticity is entering and leaving the system the circulation loss between vortice was not quantified, explaining why $\frac{d\Gamma}{dt}$ remains constant.

11. Discussion and open questions

Through the present study was seen that viscosity is who decides if vorticity can travel without diffusing itself, curl up, accumulate and form vortices. In a word is who decides how will the flow evolves. The interesting thing is that after being so influential in the flow pattern everything was in vain because the configuration of steady state regardless of the number Re (100-10,000) is very similar, a positive vortex has taken the cavity and two or three vortices were cornered. Latter observation trigger on of the most important remaining open

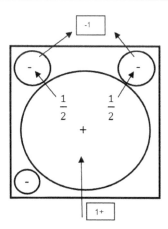

ig. 14. Vortex diagram

ᵤestion for future studies, why after so many turns, so many games, the flow reaches the ame configuration?. It is believed that a study from game theory involving two players, positive vorticity" and "negative vorticity" who fight a common good, the space of the avity, can clarify why the positive vortex end taking the whole cavity behavior that is not chieved in the deep cavity scenario. Along with the latter question, other two remain open. irst would be to answer, why the configuration of stable state coincide when the system an not store more vorticity and secondly why can not be achieved by the square cavit flow ᵣe configuration that occurs to happen in the deep cavity between the positive and negative nowing before that during the flow evolution this configuration is achieved but then lost.

2. Conclusions

ᴀmong all the results it was clearly seen the power and the preponderance of the viscosity in ᵣe evolution of cavity flows, how it affects the dynamics of vortices, transient or evolution of ᵣe flow and the accumulation or dissipation of energy. Was also observed the periodicity of ᵗeady-state flow for both cavities being the first to show a complete cycle of periodicity in the ᵉep one. In conjunction with the above the feeding channels definition were proposed which ᵥere key to understanding the transient flow. It was also proposed a transient "Bifurcation" ᵢnce they vary dramatically as the number of Re is increased. This "Bifurcation" is mainly ᵤe to viscosity.

ᴀs for deep cavities in addition to finding the periodicity of the flow for Re 8.000 it was ᵣesented an interesting phenomenon observed in Sec.5.1.2 where a *quasi cavity* is created ᵣat replicates cavity flow transients that occur before reaching steady state in a square cavity. ᵢnally, the numerical method implemented, based on the equations presented in (Chen, 2009; ᴄhen et al., 2008), was a great help for the simplicity of its programming and its primitive ᵃriable, vorticity, was central in the study.

3. Acknowledgments

ᵀhe authors are very greatful to Dr.Omar López for helpful discussions and advice.

14. References

Auteri, F., Parolini, N. & Quartapelle, L. (2002). Numerical investigation on the stabilityo: singular driven cavity flow, *Journal of Computational Physics* 183: 1–25.

Chen, S. (2009). A large-eddy-bassed lattice boltzmann model for turbulent flow simulation *Applied mathematics and computation* .

Chen, S. & Doolen, G. (1998). Lattice boltzmann method for fluid flows, *Annu. Rev. Fluid Mech* pp. 329–364.

Chen, S., Toelke, J. & Krafczyk, M. (2008). A new method for the numerical solution o: vorticity-ÂŰstreamfunction formulations, *Computational methods Appl. Mech. Engrg* (198): 367–376.

Durbin, P. & Petersson-Rief, B. (2010). *Statistical theory and modeling for turbulent flow*, Wiley UK.

Goyon, O. (1995). High-reynolds number solutions of navier-stokes equations usin₴ incremental unknowns, *Computer Methods in Applied Mechanics and Engineerin₴* 130: 319–335.

Gustafson, K. E. (1991). Four principles of vortex motion, *Society for Industrial and Appliec Mathematics* pp. 95–141.

Hou, S., Q.Zou & S.Chen (1995). Simulation of cavity flow by the lattice boltzmann method *Computational Physics* (118): 329–347.

Patil, D., Lakshmisha, K. & Rogg, B. (2006). Lattice boltzmann simulation of lid-driven flow in deep cavities, *Computers and Fluids* 35: 1116–1125.

Pope, S. (2000). *Turbulent Flows*, second edn, Cambridge Unversity Press, Cambridge U.K.

Texeira, C. (1998). Incorporation turbulence model into the lattice boltzmann method *Internationa Journal of modern Physics* (8): 1159–1175.

Toro, J. (2006). *Dinámica de fluidos con introducción a la teoría de turbulencia*, Publicacione: Uniandes, Bogotá.

Permissions

The contributors of this book come from diverse backgrounds, making this book a truly international effort. This book will bring forth new frontiers with its revolutionizing research information and detailed analysis of the nascent developments around the world.

We would like to thank Dr. Harry Edmar Schulz, Dr. Andre Luiz Andrade Simoes and Dr. Raquel Jahara Lobosco, for lending their expertise to make the book truly unique. They have played a crucial role in the development of this book. Without their invaluable contribution this book wouldn't have been possible. They have made vital efforts to compile up to date information on the varied aspects of this subject to make this book a valuable addition to the collection of many professionals and students.

This book was conceptualized with the vision of imparting up-to-date information and advanced data in this field. To ensure the same, a matchless editorial board was set up. Every individual on the board went through rigorous rounds of assessment to prove their worth. After which they invested a large part of their time researching and compiling the most relevant data for our readers. Conferences and sessions were held from time to time between the editorial board and the contributing authors to present the data in the most comprehensible form. The editorial team has worked tirelessly to provide valuable and valid information to help people across the globe.

Every chapter published in this book has been scrutinized by our experts. Their significance has been extensively debated. The topics covered herein carry significant findings which will fuel the growth of the discipline. They may even be implemented as practical applications or may be referred to as a beginning point for another development. Chapters in this book were first published by InTech; hereby published with permission under the Creative Commons Attribution License or equivalent.

The editorial board has been involved in producing this book since its inception. They have spent rigorous hours researching and exploring the diverse topics which have resulted in the successful publishing of this book. They have passed on their knowledge of decades through this book. To expedite this challenging task, the publisher supported the team at every step. A small team of assistant editors was also appointed to further simplify the editing procedure and attain best results for the readers.

Our editorial team has been hand-picked from every corner of the world. Their multi-ethnicity adds dynamic inputs to the discussions which result in innovative outcomes. These outcomes are then further discussed with the researchers and contributors who give their valuable feedback and opinion regarding the same. The feedback is then collaborated with the researches and they are edited in a comprehensive manner to aid the understanding of the subject.

Apart from the editorial board, the designing team has also invested a significant amount of their time in understanding the subject and creating the most relevant covers. They scrutinized every image to scout for the most suitable representation of the subject and create an appropriate cover for the book.

The publishing team has been involved in this book since its early stages. They were actively engaged in every process, be it collecting the data, connecting with the contributors or procuring relevant information. The team has been an ardent support to the editorial, designing and production team. Their endless efforts to recruit the best for this project, has resulted in the accomplishment of this book. They are a veteran in the field of academics and their pool of knowledge is as vast as their experience in printing. Their expertise and guidance has proved useful at every step. Their uncompromising quality standards have made this book an exceptional effort. Their encouragement from time to time has been an inspiration for everyone.

The publisher and the editorial board hope that this book will prove to be a valuable piece of knowledge for researchers, students, practitioners and scholars across the globe.

List of Contributors

Samet Y. Kadioglu
Idaho National Laboratory, Fuels Modeling and Simulation Department, Idaho Falls, USA

Dana A. Knoll
Los Alamos National Laboratory, Theoretical Division, Los Alamos, USA

O. Eichwald, M. Yousfi, O. Ducasse, N. Merbahi, J. P. Sarrette, M. Meziane and M. Benhenni
University of Toulouse, University Paul Sabatier, LAPLACE Laboratory, France

P. Domínguez-García
Dep. Física de Materiales, UNED, Senda del Rey 9, 28040. Madrid, Spain

M.A. Rubio
Dep. Física Fundamental, UNED, Senda del Rey 9, 28040. Madrid, Spain

Guiji Wang, Jianheng Zhao, Binqiang Luo and Jihao Jiang
Institute of Fluid Physics, China Academy of Engineering Physics, Mianyang City, Sichuan Province, China

Tian You Fan
Department of Physics, Beijing Institute of Technology, Beijing, China

Zhi Yi Tang
Southwest Jiaotong University Hope College, Nanchong, Sichuan, China

Hitoshi Miura
Department of Earth Planetary Materials Science, Graduate School of Science, Tohoku University, Japan

Sanjeev R. Inamdar
Laser Spectroscopy Programme, Department of Physics, Karnatak University, Dharwad, India

José Rafael Toro and Sergio Pedraza R.
Grupo de Mecánica Computacional, Universidad de Los Andes, Colombia

Printed in the USA
CPSIA information can be obtained
at www.ICGtesting.com
JSHW011416221024
72173JS00004B/558

9 781632 381361